T0177558

Rigor and Structure

John P. Burgess joined the philosophy department at Princeton University shortly after receiving his Ph.D. in logic from Berkeley, and has remained there ever since, serving for many years as director of undergraduate studies. He is the author or co-author of seven previous books and over a hundred papers and reviews in mathematical and philosophical logic, philosophy of mathematics and of language, and history of analytic philosophy.

Praise for *Rigor and Structure*

'It will be no surprise to those familiar with Burgess's other work that this book is written with exceptional clarity . . . mathematicians and philosophers alike will gain much from Burgess's specific insights about particular parts of mathematical practice; but they will also find his larger picture of modern mathematics extremely illuminating.'

Richard Pettigrew, *Philosophia Mathematica*

'*Rigor and Structure* is an enjoyable read. Given the weight of its topics, it is almost breezy . . . the book is overall quite thought provoking, insightful, and carefully written. It should not be overlooked by philosophers of mathematics who want to make sure that they are asking the questions that are most interesting to contemporary mathematics.'

Mark Zelcer, *Metascience*

'Burgess knows his subject matter well, and his clear and engaging writing makes his prose a pleasure to read.'

A. C. Paseau, *British Journal for the Philosophy of Science*

'This book is a delight to read . . . As a pioneering work in this new field, *Rigor and Structure* makes a solid impression.'

Toby Meadows, *Australasian Journal of Philosophy*

Rigor and Structure

John P. Burgess

OXFORD
UNIVERSITY PRESS

OXFORD

UNIVERSITY PRESS

Great Clarendon Street, Oxford, OX2 6DP,
United Kingdom

Oxford University Press is a department of the University of Oxford.
It furthers the University's objective of excellence in research, scholarship,
and education by publishing worldwide. Oxford is a registered trade mark of
Oxford University Press in the UK and in certain other countries

© John P. Burgess 2015

The moral rights of the author have been asserted

First published 2015
First published in paperback 2020

Published in the United States of America by Oxford University Press
198 Madison Avenue, New York, NY 10016, United States of America

British Library Cataloguing in Publication Data
Data available

Library of Congress Cataloging in Publication Data
Data available

ISBN 978-0-19-872222-9 (Hbk.)
ISBN 978-0-19-882267-7 (Pbk.)

Preface

Few buzzwords are more often encountered in philosophical discussions of the nature of mathematics than "rigor" or "structure," and few are more diversely understood. Thus while we are commonly told that the distinctive method of mathematics is rigorous proof, and that the special topic of mathematics is abstract structure, even a superficial look at the literature in which such formulations appear quickly shows that neither mathematicians nor logicians nor philosophers are agreed as to the exact sense in which mathematics is or ought to be concerned with rigor, or with structure. Differing perspectives on such issues can be of practical as well as theoretical importance, leading to divergent views on matters of policy: on the role computer-dependent proofs should be allowed to play, or the relative importance of set theory as opposed to category theory. The aim of the present essay is to clarify our understanding of mathematical rigor and mathematical structure, and especially of the relation between the two, and to sketch how such a clarified understanding can bear on disputed policy questions.

An account of the origin of this work may help explain its organization. Much writing by philosophers about mathematics in recent decades has been concerned with two views about the nature of mathematical objects and mathematical existence, known in the literature as *nominalism* and *structuralism*. Fifteen and more years ago, my colleague Gideon Rosen and I wrote a study of nominalism (Burgess and Rosen 1997). Even at the time I had the feeling that a similar study of structuralism was called for, but other projects intervened, so that only five or so years ago did I begin serious work. My developing views on structuralism were presented in lectures delivered under various titles at Stanford, NYU, Penn, Bristol, Paris, Oxford, and at home in Princeton, and one that would have been delivered in Helsinki but for Eyjafjallajökull. The thesis I was led to was that *the features of mathematical practice that structuralists want to explain in terms of the peculiar nature of mathematical objects are better explained in a different way, as consequences of the way the ancient ideal of rigor is realized in modern mathematics.* More specifically, what that way of working involves is that *the individual mathematician is responsible for the rigor of the derivation of any new results claimed from results in the previous literature, but may remain indifferent to just how the results in the previous literature were derived from first principles.* The exposition and defense of this thesis is to be found in Chapter 3 of the present work.

When I began to work on the present book, my aim was to develop the material in my various lectures into a long paper on the thesis I have just adumbrated. However, when I lectured on my work in progress, I found that in the question-and-answer sessions afterwards, members of the audience always wanted to press a question I had passed over, namely, the question of what the first principles from which the rigorous development of mathematics proceeds should be understood to be. More specifically, the question raised was whether set theory or category theory should be thought of as the starting point. When I was presented with the opportunity to co-teach a graduate seminar on set theory versus category theory, I eagerly accepted, resolving in preparation for the task to think harder about the question of first principles, which I had been inclined to postpone. I was eventually led to a corollary to my main thesis, which in hindsight may appear an obvious consequence of it: *The working mathematician may remain indifferent not only to how results in the existing literature were derived from first principles, but also to what exactly those first principles are—though they* might as well be *those of standard set theory.* The exposition and defense of this corollary is to be found in Chapter 4 of the present work.

As my preparation for the seminar began, it became clear to me that before launching into making the case for my thesis I had better begin, by way of background, with some general discussion of rigor. At first, I expected that some brief prefatory remarks would suffice, but as I went deeper into the matter I found there was more that needed explaining about the ideal of rigor, about how it is realized in present-day mathematics, and about how it came to be realized in that way, than I had originally anticipated. Such preliminary explanations are now to be found in Chapter 1 of the present work.

As my preparation continued, it also became clear to me that if I were to discuss category theory versus set theory, and the corollary to my main thesis, I would need to explain not only what rigor is, but also how set theory came to occupy the position some category theorists would now like to challenge. I mean the position of being in some sense the accepted foundation or starting point for rigorously building up the rest of mathematics. Such further preliminary explanations are given in Chapter 2 of the present work.

Thus, what I originally expected to be a longish paper has turned into the not-so-short book I now lay before the reader. Lest the prospective reader be daunted by the sheer number of pages to follow, let me add an explanatory remark. There is an old, and doubtless false, story about the mathematician Euler confronting the philosopher Diderot at the court of Catherine the Great with a nonsensical mathematical refutation of atheism: "Sir, $(a + b)^n/n = x$, hence God exists. Reply!" I dislike this kind of bullying by invoking technicalities, and have accordingly tried in the present work to keep all mathematics at as elementary a level as I knew how, for the sake of philosophical readers. Equally, I have tried to explain all philosophical background in as elementary and jargon-free way as I knew how, for the sake of mathematical readers. Such is and has long been the state of lack of communication between the two fields that no other course seemed to me feasible, if I wanted to have any hope of not utterly putting off readers of one sort or the other. The result is that mathematical readers will find a fair amount of elementary mathematical exposition intended for philosophers that they will want to skip or skim over, while philosophical readers will similarly find a fair amount of elementary philosophical exposition intended for mathematicians that they will want to read over quickly, if they read it over at all. Thus, for either class of readers the present work will effectively be shorter than it appears to be.

Princeton, November 2014

Acknowledgments

I now turn to the pleasant duty of acknowledging intellectual debts. My oldest and deepest are to two of my teachers: Arnold E. Ross, in whose summer program in number theory for high-school students I first learned what a proof is, and Ivo Thomas, from whom I learned in the same program the elements of classical and intuitionistic formal logic. I am also much indebted to Stewart Shapiro, once a fellow student with me of Ross and Thomas, and now the structuralist philosopher whose work has most concerned me. He was present and participated in the discussion at more than one of my talks on structuralism, and besides engaged in an informative correspondence with me about some of the issues. I owe much also to other questioners from the audiences after my talks, and to two other structuralist philosophers with whom I have been in communication over the years, Geoffrey Hellman and Charles Parsons.

Closer to home, to Hans Halvorson, the Princeton colleague with whom I jointly conducted the graduate seminar on set theory and category theory, and later an undergraduate seminar on philosophy of mathematics, I owe among other things much of such knowledge of category theory as I possess. I must hasten to emphasize, however, that he is not in any way responsible for the opinions I express, and especially not for any that may be uncongenial to his friends among category enthusiasts. I also profited from attending my colleague Benjamin Morison's seminar on Euclid's *Elements*, book 1, and to a greater degree than will be apparent from my scattered remarks on Euclid in this work.

I received useful comments on earlier drafts from Oystein Linnebo, from my emeritus colleague Paul Benacerraf, from my student Jack Woods, from my friends Juliette Kennedy and Jouko Väänänen, and on typographical matters from Michael Scanlan. I am grateful also to OUP's two anonymous referees, who convinced me that some of the things I was inclined to say were open to objections, and others to misinterpretations, which I had not anticipated. The many members of the staff of OUP and affiliates who became involved with the project were exceptionally efficient and helpful throughout, from the first inception of the idea for such a volume to the final proofreading.

Contents

1

Rigor and Rigorization

What Is Rigor?

Go to any university biology library. Pick at random two or three recent volumes of different mainstream journals. Sit down and flip through the pages of several articles in each, not hoping to absorb their substantive content, but merely observing their gross form. Now go over to the mathematics library, and repeat the process. The reader who performs this exercise cannot fail to be struck by an enormous difference in format between papers in a natural science like biology and papers in mathematics. In biology one finds descriptions of experimental design and statistical analyses of data collected; in mathematics, a series of definitions, theorems, and proofs. The distinctive definition–theorem–proof format of professional publications is the single most conspicuous feature of mathematical practice.

In any science, a published paper may turn out to be less innovative or less insightful than it appeared to be to the editors and referees responsible for its acceptance by the journal in which it appears. In such a case, there will simply be a dearth of subsequent citations of the paper, which will fall into oblivion. It may also happen, though less frequently, that there is something so wrong with the paper that professional ethics would call for the publication of a retraction, whether or not one actually appears. In biology, some error in data collection or analysis may be uncovered. In mathematics, some flaw or gap in a proof may be recognized, so that it has to be acknowledged that what appeared to be a proof was not really a proof. In some cases of this kind, the retraction may offer a counter-example to what was originally published as a "theorem"; in others, it may merely be indicated that the previously published result must now be regarded as a mere "conjecture."

The quality whose presence in a purported proof makes it a genuine proof by present-day journal standards, and whose absence makes the proof spurious in a way that if discovered will call for retraction, is called *rigor*. Or at least, that is one of several uses of that term, and the one that will be of concern here. Now the definition of rigor just given, as that which is demanded of proofs in contemporary mathematics, identifies rigor by the role it plays in the mathematical enterprise; but what one would like to have, and what we will be seeking in the pages to follow, is a more intrinsic characterization: an account of what rigor, thus identified, amounts to.

In examining this question, it is natural to begin with what mathematicians themselves may have to say about the matter. They are the experts, after all. For mathematicians, the ability to judge an argument's rigor in our sense, to evaluate whether a proof is up to the standard required in professional publications, is a crucial skill. Mathematicians need to be able to evaluate the rigor of their own work before submitting it, and to judge the rigor of the work of others when called on to serve as referees.

And mathematicians are generally quite able in such matters. Evaluating the rigor of a journal submission, though often tedious in the way that proofreading for typographical errors is tedious, may be the most routine part of refereeing, much more cut and dried than evaluating how innovative or insightful a paper is. For rigor is only a preliminary qualification for publication, as lacking a criminal record may be a preliminary qualification for certain kinds of employment. Failure to meet the requirement may mean rejection, but success in meeting it does not necessarily mean acceptance.

Mathematicians learn what rigor requires, learn to distinguish genuine from bogus proofs, during their apprenticeship as students. This skill is acquired mainly by observation and practice, by example and imitation. Typically, the student "learns what a proof is" in connection with a course on some substantive mathematical topic, perhaps "arithmetic" in the sense of higher-number theory, or more often "analysis" in the sense of the branch of mathematics that begins with calculus. The assessment of purported proofs for rigor is generally *not* the topic of a separate course. In particular, it is *not* generally learned by studying formal logic, courses which tend to be taken more by students of philosophy or computer science than of mathematics.

Mastery of a practical skill does not automatically bring with it an ability to give a theoretical account thereof. Even a champion racing cyclist may find it difficult to explain how to ride a bike, if tied hand and foot and required to indicate what to do using words only, without any physical demonstration. In the same way, expert mathematicians are not automatically equipped to give an accurate theoretical description of the nature of rigorous proof. In any case, they nowadays very seldom attempt to do so. Most mathematicians seem to stand to rigor as Justice Stewart stood to obscenity: They can't define it, though they know it when they see it.

Mathematicians' views on the nature of rigor and proof may more often be expressed in aphorisms and epigrams, in anecdotes and jokes, for instance, of the common "an engineer, a physicist, and a mathematician . . ." genre,[1] than in formal theoretical pronouncements. Browsing on the shelves of the library, or perhaps more conveniently using a search engine, will turn up many crisp sayings and droll stories from mathematicians present and past, major and minor, pure and applied. Most of the sayings are semi-jocose and not to be taken completely literally; many if so taken would be rather obviously false: Though there may be a grain of truth in them, considerable threshing and winnowing may be needed to separate it from the chaff. All together, this material represents something less than a full-blown theoretical account, and though mathematicians' witticisms and humor about rigor and proof contain many pointers, these don't all point in the same direction.

To give at least a couple of examples, a search will turn up many versions, variously formulated and variously attributed, of the dictum whose simplest form is just "a proof is what convinces." (A couple of more refined versions will be taken up later.) A search may also turn up different retellings from different authors of the story about the young gentleman and his tutor. In case the reader is unfamiliar with this tale, here is the version, set in eighteenth-century France, given by Isaac Asimov:

[A] poor scholar was hired to teach geometry to the scion of one of the nation's dukedoms. Painstakingly, the scholar put the young nobleman through one of the very early theorems of Euclid, but at every pause, the young man smiled amiably and said, "My good man, I do not follow you." Sighing, the scholar made the matter simpler, went more slowly, used more basic words, but still the young nobleman

---

[1] The oldest may be the following. A mathematician, a physicist, and an engineer travel on a train in Scotland. They catch a glimpse of a sheep. "Sheep in Scotland are black," says the engineer. "There is at least one black sheep in Scotland," says the physicist. "There is at least one sheep in Scotland with at least one black side," says the mathematician.

said, "My good man, I do not follow you." In despair, the scholar finally moaned, "Oh, monseigneur, I give you my word what I say is so." Whereupon the nobleman rose to his feet, bowed politely, and answered, "But why, then, did you not say so at once so that we might pass on to the next theorem? If it is a matter of your word, I would not dream of doubting you." (Asimov 1971, 456)

The unstated moral of the story is precisely that, contrary to the dictum that a proof is what convinces, *not* everything that convinces is a proof.

If several Fields medalists, often described as the nearest mathematical analogue of Nobel laureates, were to pledge their word of honor that a certain mathematical conjecture is true, few of us would dream of doubting them. But publication of an assemblage of such testimonials in favor of some outstanding conjecture, say the one known as the *Riemann hypothesis*, though it might well produce widespread conviction that the conjecture is true, would not qualify for the million-dollar prize offered by the Clay Foundation for the solution of one of its "Millennium Problems" (see Carlson et al. 2006). For to qualify for the prize one must publish, in a peer-reviewed journal, a *proof*, and whatever exactly is needed for a proof, a collection of testimonials is not enough. So at the very least, conviction derived from testimony must be distinguished from conviction derived from proof: The supporting arguments for a piece of rigorous mathematics cannot just be appeals to authority, or tradition, or revelation, or faith. The further clarification of the nature of rigor will require characterizing what more is excluded.

Before launching into that project, it should be mentioned that the dictum that "a proof is what convinces" sometimes fails for the opposite of the reason we have been considering so far: Genuine proofs do not always convince, at least not on first exposure. Anyone who has taught undergraduate mathematics knows that proofs do not always immediately convince students, but indeed they do not always immediately convince even the proof's own author. The skeptical Scottish philosopher David Hume put it well:

There is no ... Mathematician so expert in his science, as to place entire confidence in any truth immediately upon his discovery of it, or regard it as any thing, but a mere probability. Every time he runs over his proofs, his confidence increases; but still more by the approbation of his friends; and is rais'd to its utmost perfection by the universal assent and applauses of the learned world. (Hume 1739, part IV, section I)

Such are the effects of an awareness of human fallibility.

Aristotle and Logic

If we cannot quite find the theoretical characterizations of proof and rigor
that we want in the remarks of mathematicians, perhaps we will do better if we turn to logicians. And where better to begin than with Aristotle,
the very first logician? His *Prior Analytics* (Cooke and Tredennick 1938)
launched logic by launching the theory of *syllogisms*, a dozen or so forms
of argument exemplified by the pattern of reasoning the medievals called
Barbara:

All Bs are Cs.
All As are Bs.
Therefore, all As are Cs.

It is only a slight exaggeration to say that logic made no permanent
advance beyond the point reached in this work (every step forward
being followed by a step back) until the time of Gottlob Frege in the later
nineteenth century.

It is undeniable that the theory of syllogisms is more impressive
when considered as the work of a single individual, starting from essentially nothing, over the course of a career devoted to many other intellectual pursuits, than when considered as the best the human race could
produce in the domain of logic in two thousand years of activity. There
are undeniable defects in Aristotle's pioneering work, and the great
twentieth-century logician Bertrand Russell treats them rather severely
in his chapter on Aristotelian logic in his *History of Western Philosophy*
(Russell 1946, ch. 22). One defect noted by Russell is an overestimation
of syllogism as compared to other forms of deduction, and another is an
overestimation of deduction as compared with other forms of reasoning.

These defects matter when we come to Aristotle's *Posterior Analytics*
(Tredennick and Forster 1960), the work in the Aristotelian corpus most
relevant to our present concerns. On account of the overestimation of
deduction, Aristotle there does not speak of what mathematical proof in
particular requires, but rather of what scientific demonstration in general
requires. If we take "scientific" in its modern sense, this is a definite error,
for our starting point was precisely the observation that mathematics is
unlike other sciences. On account of the overestimation of the syllogism,
Aristotle does not speak of proceeding deductively, but rather of proceeding syllogistically. This is also a definite error since, as some ancient

logicians already objected, and as any post-Fregean logician would agree, mathematics requires deductions that cannot be cast into the form of syllogisms.

Correcting for these two features, we can, however, read in—or read into—Aristotle a statement of what remains today the requirement of rigor by which the individual mathematical author hoping to achieve publication in a recognized journal is confronted:

> (1) Mathematical rigor requires that every new proposition must be deduced from previously established propositions (coming either from earlier in the same paper, or from the earlier literature).

Hence the abundance of external citations and internal cross-references in mathematical papers (and hence the numbering of theorems, needed for purposes of cross-reference). Hence also the ubiquitous appearance in proofs in journal articles of the words and phrases typically used to signal deductive steps: "whence," "hence," "thence," "therefore," and so on.

Less application of the principle of charity is needed to find another important observation in Aristotle. It would be rather pointless for the individual mathematician to observe the requirements of rigor if the previous literature cited did not itself do so. Rigor really only has a point if it is observed by a whole community, and is to be found not just in the work of mathematicians as individuals, but in the corps of mathematicians collectively, in the science of mathematics as a whole. And Aristotle addresses precisely this question of rigor across the whole discipline, bringing out the following point:

> (2) On pain of circularity or infinite regress, if later propositions must be proved from earlier ones, then we must start from some unproved propositions or *postulates*.

Such unproved propositions are also called *axioms*.

Besides issues about deduction, Aristotle addresses issues about definition, but he understands by the term traditionally so translated something rather different from what we understand by it today.[2] As to definition in

---

[2] Aristotle himself reserves the term traditionally translated as "postulates" for the unproved propositions peculiar to a given subject, as opposed to those axioms that are "common notions" shared with other subjects. And for him, a "definition" is something more than a statement of the meaning of a word: something aspiring to characterize the "essence" of a thing.

its modern sense, modern mathematics may be said to impose requirements exactly parallel to points (1) and (2), namely the following:

(3) Mathematical rigor requires that every new notion must be defined
in terms of previously explained notions (either from earlier in the
same paper, or from the earlier literature).

(4) On pain of circularity or infinite regress, if later notions must be
explained in terms of earlier ones, then we must start from some
unexplained notions or *primitives*.

And that is essentially all that rigor as such and in itself requires of
definition.

Of course, when it is a question, not of mathematics done rigorously
from the start, but of rigorous reconstruction of pre-existing non-rigorous
mathematical thinking, definitions also need to be faithful, so far as is
compatible with the ideal of rigor, to pre-existing usages, after the manner of lexicographers' definitions. And of course, a definition, though perfectly rigorous, will be pointless unless something useful can be done with
the defined notions: Recall that rigor is only a preliminary requirement.

Note that the requirements of rigor as formulated in points (1)–(4) pertains entirely to how new mathematical material (notions and results) is
derived proximately from older material, and ultimately from first principles (primitives and postulates). It is about what one does with the first
principles once one has them, and not about where one gets the principles
to begin with. A long tradition extending from ancient to modern times
wished to impose a requirement along the following lines:

(5) The meaning of the primitives and the truth of the postulates must
be evident.

It would be worse than a mere oversimplification to attribute requirement
(5) to Aristotle himself, but I need not go into the matter, for a double reason: Not only is no restriction on first principles part of the notion of rigor
as such, which is our topic here, but also, as will be seen later, modern
mathematics has definitely abandoned any such requirement as (5), even
as an ideal honored in principle, let alone as a requirement enforced in
practice.

The account of the ideal of rigor given so far is incomplete, not in
omitting restrictions on what may be taken as first principles, but rather
in requiring supplementation by an account of what definition and

deduction, the processes that are supposed to take us in stages from first principles to the latest notions and results, amount to. But it is better to postpone offering such an account than to offer a premature and inadequate one.

The route I propose to take is to seek a gradual clarification of the nature of deduction as it figures in mathematical proof, at first just by contrasting it with types of argumentation that clearly are not deductive. In so doing, among other things I will be massively reinforcing the moral of the nobleman-and-scholar story, that not everything that convinces is a proof, at least not by the rigorous standards of proof enforced by current journals: Not only may what convinces the non-expert be simply testimony by an expert, but also, and more importantly, what convinces the experts may be a heuristic argument rather than a rigorous proof. In the end, at least a half-dozen types of reasoning other than reliance on testimony will be found to be excluded by the requirement of rigor.

Discovery and Justification

One of the first things to note about rigor is that it pertains only to what philosophers of science call the "context of justification," as opposed to the "context of discovery." Theorems may be discovered as conjectures years or decades, or even centuries (as in the case of the famous Fermat result) before proofs are found for them. And when a proof is discovered, in no interesting case is it discovered by first discovering the first step, second discovering the second step, and so on. Often, however, all traces of the discovery process are effaced in writing up a result for journal publication. "A cathedral is not a cathedral," the great Carl Friedrich Gauss is reported to have said (see Bell 1956, 305), "until the last scaffold is down and out of sight." But moving away from the shelves with the current journals, there are works in the mathematics library where one can find information about mathematical discovery.

One classic of this kind is Jacques Hadamard's *Psychology of Invention in the Mathematical Field* (1945). Describing the mentation of leaders at the very top of the mathematics profession, Hadamard emphasizes the role of *nonverbal thought*, of thinking in images that resists statement in words. Hadamard also emphasizes the role of *sudden insight*, flashes of inspiration that may suddenly erupt from the unconscious when, after long rumination over a problem, conscious pondering of the problem, is for a time set aside. A quite different classic is George Polya's *How to Solve It* (1945), a book of advice for students. Polya emphasizes the role in discovery of *inductive generalization*, projection of a general result from a series of special cases individually verified by computation or otherwise,[3] and the closely related process of *analogical extrapolation*, where one considers what would be the most natural parallel to a known result in a new situation. (He eventually elaborated his study of such matters into a two-volume work, Polya 1954.)

Now writers like Hadamard and Polya, both distinguished mathematicians, would doubtless insist that whatever is discovered by nonverbal thought, or sudden insight, or inductive generalization, or analogical extrapolation needs to be justified by rigorous proof; but what is important for our present topic in descriptions of discovery processes in mathematics is that such descriptions as Hadamard's and Polya's help us to imagine what a non-rigorous mathematics, sharply contrasting with what is to be found in recent journal articles in the library, might look like.

Actually, we can imagine *two* forms of non-rigorous mathematics. One form, *mathematics without proofs*, would simply dispense with the definition–theorem–proof format altogether. I do not mean merely that proofs would be omitted in the write-up of results. That kind of omission of proof can be seen already in the current literature. Survey-expository articles in the *Bulletin of the American Mathematical Society* (*BAMS*), for instance, include at most outlines or comments on proofs not presented, and many compendia and handbooks of mathematical formulas and results, works on the order of the classic synopsis Carr 1886, contain only minimal information about how the formulas they contain are derived or justified. I mean, rather, that mathematics might be done without any proofs *even offstage*. Results might simply be announced as the product of thinking in images that cannot be put into words or of flashes of inspiration perhaps sprung from the unconscious, without offering further justification or explanation.

Another form of non-rigorous mathematics, *proofs without rigor*, would still contain general assertions that might be *called* "theorems," and

---

[3] Notice that what is at issue is *not* so-called "mathematical induction," where one establishes that a result holds for zero, and that it holds for the successor of any natural number for which holds, and concludes that it holds for all natural numbers. *That* is a form of rigorous proof, not heuristic argumentation.

supporting arguments that might be *called* "proofs," though according to today's professional standards the former would really only be *conjectured* results and the latter only *heuristic* considerations. The supporting arguments might, for instance, involve such forms of non-demonstrative reasoning as inference from particular examples to general patterns or by comparison between a known case and an unknown case.

But it is not really necessary to exercise the imagination in order to form a picture of what non-rigorous mathematics might be like: One can actually see examples of something like it, and some of them in works likely to be found on the shelves of a mathematics library.

A notable apparent case of mathematics without proofs is provided by the letter, full of strange and wonderful formulas, by which the then-unknown Indian prodigy Srinivasa Ramanujan introduced himself to the well-known English don G. H. Hardy.[4] Ramanujan was almost entirely self-taught, and mainly from works like the Carr volume just mentioned, and so had had little exposure to proofs at the time he wrote, and few are included in his extensive notebooks.[5]

More important for our purposes than the case of mathematics without proofs is the case of proofs without rigor. For these, published sources are much more abundant, especially in applied work. This is most especially so if one looks beyond the work of people called "applied mathematicians" rather than "pure mathematicians"—people whose work is represented in the many journals of the Society of Industrial and Applied Mathematicians—to the work of people not called "mathematicians" at all, and especially of those called "physicists." Gone are the days of Newton when a single individual could be to an equal degree of eminence an experimental and a theoretical physicist, and an applied and a pure mathematician. Today the prevailing division of intellectual labor is such that one must distinguish even the role of a "mathematical physicist," someone like Roger Penrose, a mathematician much concerned with physics, from the role of a "theoretical physicist," someone like Edward Witten, a physicist much concerned with mathematics. And a first crucial difference is

[4] For background, see Newman 1956b, which incorporates excerpts from the letter to Hardy, and from Hardy's own reconstruction, in a later memoir, of his immediate reaction to it.
[5] He even reported that some of his formulas were revealed to him in dreams. But Bruce Berndt, who edited and annotated Ramanujan's notebooks for publication, suggests (1998, 4) that in many cases he had worked out a proof of sorts on a slate, but recorded only the result in his notebooks.

just this, that professional colleagues expect rigor from the former and not from the latter.

Even if one restricts one's attention to mathematics rather than physics, and pure rather than applied mathematics, one will find on the shelves of the library at least one periodical, *Experimental Mathematics*, devoted to non-rigorous material. This journal is especially known for publishing papers presenting conjectures suggested by computer simulations or calculations, and not (yet) having rigorous proofs. Hence the not entirely respectful nickname, *Journal of Unproved Theorems*. According to its editors' official policy statement on the inside cover of each issue, they publish "original papers featuring formal results inspired by experimentation, conjectures suggested by experiments, and data supporting significant hypotheses."

There is this limitation to the theoretical physics literature as a source of examples of proofs without rigor, that the non-rigorous mathematicizing in that literature often occurs in discussions leading from one mathematically formulated physical hypothesis to another, in which purely mathematical considerations are not sharply separated from substantive physical considerations.[6] Thus one may not actually see much in the way of the isolation of purely mathematical results presented as "theorems" (perhaps not quite rigorously formulated) and supported by purely mathematical arguments presented as "proofs" (not amounting to rigorous deductions).

There is a limitation also to *Experimental Mathematics* as a source of examples. For that journal, being the product of an era in which

---

[6] A fairly accessible example is the derivation in Einstein's great 1905 paper on Brownian motion of an equation expressing Avogadro's number, the number of molecules in a "mole" of a substance (about a tablespoon of water, for instance), in terms of quantities potentially measurable using a microscope. An analysis is given in Pais 1982, part II, sect. 5. The equation is obtained by cancellation from two equations for a diffusion process (for instance, of dye in water) that are, as Pais notes, *based on two incompatible models*, discrete and continuous. The case perhaps illustrates Einstein's own assertion (1949, 683–4) that to a systematic epistemologist the working scientist must appear an unscrupulous opportunist. The two models being incompatible, perfect rigor in the mathematical derivation of the two equations from the two models would be rather beside the point, would it not? At any rate, none is on offer. As Pais emphasizes, given the features of the Einstein derivation just mentioned, it was not just Jean Perrin's subsequent experimental measurement of the relevant quantities that convinced the last scientific doubters of the reality of atoms, but the fact that the estimate for Avogadro's number obtained in this way, and other estimates that were obtained by Einstein and others, generally in a fashion not conspicuously more rigorous, using a variety of other equations, pertaining to a variety of other physical phenomena, all produced approximately equal figures.

rigorous mathematics predominates, exemplifies only a very "tame" kind of non-rigor: It is always carefully indicated what has been rigorously established and what has only been experimentally confirmed, so that there is no danger of confusing theorem with conjecture, or heuristics with proof.

The best source for mathematics that does not conform to present-day standards of rigor, very abundantly available in even the smallest college library, containing statements of many important theorems of pure mathematics, and often quite wild rather than tame in mixing rigorous and non-rigorous steps together haphazardly, is to be found, not in any contemporary journal, but in historical sources. For this reason among others, philosophers' discussions of the role of rigor and proof in mathematics often turn to historical examples, as indeed do philosophical discussions of any number of other aspects of mathematical practice.

The number of historical episodes that get discussed or alluded to in this way is large, and any one of them could be, and any number of them have been, made the topic of a monograph as long as the present book or longer.[7] Since my aim here will be to work in presentation of a large sampling of historical examples of the kind that tend to be cited by philosophers, while keeping my a discussion limited in overall length—I am writing an informal essay, not a systematic treatise—and since I further wish to presuppose knowledge of little more than high-school or at worse freshman mathematics, except in occasional parenthetical remarks or footnotes for the *cognoscenti*, I must inevitably abridge and condense a good deal, besides relying for the most part on secondary rather than primary sources.

In the pages immediately following, I will simply be recalling selected key aspects of selected key episodes of the relevant history. I will be treating the mathematics at a semipopular rather than a fully technical level, with occasional references to other sources for fuller popular accounts and citations of technical accounts.[8] And I will be

---

[7] One of the earliest and best-known examples of a philosophical monograph on an historical case study is Lakatos 1976. This celebrated account remains eminently readable, despite being rather heavy on philosophical editorializing. It concerns the fascinating, but perhaps somewhat atypical, situation of the "Euler characteristic."

[8] For most developments, information is readily available either in Rademacher and Toeplitz 1966 or in selections from Newman 1956c.

treating the history in a summary fashion, again referring the reader elsewhere, and to begin with to standard general narrative histories, for more information and citations of both primary sources and of specialist studies.[9] My approach to history, besides being abridged and condensed, will be unashamedly "Whig", emphasizing the aspects of the past that seem most important to us today, which are not always those that seemed most important at the time. I will also adopt an old-fashioned "internalist" rather than a more fashionable "externalist" orientation, and emphasize internal mathematical or scientific reasons for changes in mathematical practice over time rather than external social pressures.[10]

[9] For most developments, the best-known large-scale history of mathematics, Kline 1972, will more than suffice for present purposes, for all that present-day historians would see a need for revision on certain points. Bibliographic data for mathematical works of times past alluded to in passing in this chapter can be found in Kline, and I will generally give such data here only in cases where I make a direct quotation from such a source. While Kline's is a work by a mathematician and largely for mathematicians, the first thirteen chapters, at least, which take us from Mesopotamia and Egypt to arithmetic and algebra in the sixteenth and seventeenth centuries, and cover all the topics from that period to which I will allude and much more also, can be read with only a knowledge of secondary-school mathematics. The mathematical prerequisites of the later parts of the book vary from chapter to chapter. The classic Smith 1929 contains (in English translation, and in some cases with modernized notation) at least excerpts from almost all the original works of writers from the sixteenth to the early twentieth centuries to which I allude. A great deal is to be learned even by the reader who has not the time to study this material in detail, just from taking a quick glance at any of its selections in the sourcebook and comparing its style to that of modern works by and about mathematics.

[10] My many citations will amount to a background reading list for the student of philosophy of mathematics almost wholly disjoint from the more formal philosophical discussions collected in such classics as Benacerraf and Putnam 1983. Along with the work of Rademacher and Toeplitz, Newman, Kline, Smith, and similar books—for there are more high-quality books of these kinds than I have space to cite—I would mention also a periodical, *The Mathematical Intelligencer*, almost every issue of which contains an article or two providing the philosopher with food for thought. Moreover, web-browsing on the keywords connected with almost any of the illustrative cases I mention will lead to even more material of the same kind that I will *not* have time to get around to. (For instance, the reader who follows up on my mention in these pages of the "Mertens conjecture" will then encounter the "Polya conjecture," the "Collatz conjecture," and others, all significant cases equally worth pondering.) I must offer one caveat, however. An ability will be required to pass over in charitable silence certain kinds of digressions into amateur philosophizing, and especially into attempts to construe mathematical objects as somehow mental entities, which anyone who has read and understood Frege knows is hopeless. It should go without saying that the semipopular works I cite are intended to supplement, not to supplant, the study of such philosophical classics as are collected by Benacerraf and Putnam.

Pre-Modern Mathematics

Among the oldest written records of mathematical thought are certain Egyptian manuscripts, for instance, the document known as the Rhind papyrus.[11] Inspection of such material shows us a form of mathematics very different from that of modern journals or textbooks. There is nothing to be seen but worked problems. These are intended to convey general methods: One can do similar problems by going through analogous steps starting from different data. But the language needed to state general propositions is not available. Truly efficient symbolic notation for algebra dates only from the sixteenth century,[12] but what the scribe who produced the Rhind papyrus, one Ahmes, and his colleagues lacked was not just an efficient symbolic notation, but even the ability to state general results in words.

From the sequence of civilizations in Mesopotamia, conveniently if inaccurately called collectively "Babylonian" after the last and greatest of them, a great deal more mathematical material survives, since clay tablets are much less perishable than papyrus scrolls. The Babylonian material, which includes extensive applications of mathematics to astronomy, is of a more varied character, but still much the same story holds for the Babylonians as for the Egyptians: We do not find anything to remind us of present-day journals. It is one thing to know how to solve problems that we today would express by quadratic equations, but quite another to have the resources to state the quadratic formula explicitly. Egyptian and Babylonian sources alike present us not just with mathematics without proofs, but mathematics without explicitly stated general theorems.

Greek mathematics arose from that of the ancient Near East. This fact is acknowledged in Greek sources in two ways, indirectly by claims to the effect that the legendary figures identified as their first mathematicians, Thales and Pythagoras, traveled and studied in Egypt and elsewhere, and directly by more explicit statements. Thus Proclus says that arithmetic was

[11] Newman 1956a provides a readily available popular account, illustrated by extracts in transcription and translation and by photographs of the original document. For a fuller picture of Egyptian mathematics on a popular level, see Reimer 2014.

[12] Trying to read anything from before the time of François Viète, whether "rhetorical algebra," where results are written out in words only, or "syncopated algebra," where results are written out using abbreviations, or for that matter early "symbolic algebra," is like trying to read, as a speaker of modern English, Chaucer (very difficult) or *Beowulf* (impossible).

first developed by the Phoenicians in connection with trade and commerce, and that geometry was first developed by the Egyptians in connection with surveying (an origin reflected in the etymology of the word "geometry", from *geo* or *earth* plus *metreia* or *measurement*). This Egyptian origin of geometry, especially, is repeatedly mentioned by the Greeks, though different writers put different spins on the fact, in a way that may tell us as much about the Greek writers as about the Egyptians they are writing of: Herodotus emphasizes the practical, citing the need to reestablish boundaries after the annual flooding of the Nile; Aristotle emphasizes the theoretical, and the emergence of a leisured priestly class with time for intellectual pursuits.

What the Greek sources do *not* say is that the kind of rigorous geometry found in Euclid and Archimedes and Apollonius came to Greece from elsewhere. On the contrary, the monumentally important idea of organizing mathematics into a deductive science, such as we find in the works of the big three just named, is the chief specifically Greek innovation. We will return to look at this famous and distinctive part of Greek mathematics later, but at this point what I want to emphasize is that, besides the rigorous geometric tradition, Greek mathematics had a less famous, less rigorous, more algebraic side, more continuous with the Near Eastern background, represented by such figures as Diophantus and Heron.

To give at least one example, here is a sample passage from the last-named figure:

We shall now inquire into the method of extracting the cube root of 100. Take the nearest cube in excess of 100 and also the nearest which is deficient; they are 125 and 64. The excess of the former is 25, the deficiency of the latter 36. Now multiply 36 by 5, the result is 180; and adding 100 gives 280. Dividing 180 by 280 gives $^9/_{14}$. Add this to the side of the lesser cube, that is, to 4, and the result is $4\,^9/_{14}$. This is the closest approximation to the cube root of 100. (Thomas 1939, 60–3)

The style here is almost Egyptian or Babylonian: a single example is worked, evidently for imitation in other cases. We today might describe the method as a kind of program or algorithm, thus:

1. Given a target number not itself a perfect cube (100), find the greatest number whose cube is less than the target number (4).
2. Note the cube of the result of step 1 ($4^3 = 64$).

3. Compute the difference between the result of step 2 and the target number (100 − 64 = 36).
4. Find also the least number whose cube is greater than the target number (5); it will be one more than the result of step 1.
5. Note the cube of the result of step 4 (5^3 = 125).
6. Compute the difference between the target number and the result of step 5 (125 − 100 = 25).
7. Take the square root of the result of step 6 ($\sqrt{25}$ = 5) (or take an approximation thereto if the result of step 6 is not a perfect square).
8. Multiply the result of step 3 by the result of step 7 (36 · 5 = 180).
9. Add the target number to the result of step 8 (180 + 100 = 280).
10. Divide the result of step 9 by the result of step 8 (180/280 = $4\,^9/_{14}$), to obtain an approximation to the cube root of the target number.

Heron is not even this explicit—step 7, in particular, is only very obscurely hinted at in the quoted passage—and certainly does not summarize the algorithm as a modern writer might, in a single formula in compact algebraic notation. The editor/translator Thomas gives the implicit formula (1939, 62 n. *a*), but in doing so he says that it is unlikely Heron himself worked with it, and the "unlikely" here is an understatement. In any case, even if he had given the formula, Heron had nothing at all to offer resembling a proof that the formula always works. Indeed, he gives no very clear statement of what "working" would consist in: How good does an approximation have to be before it counts as a "close," let alone as the "closest," approximation?

The chief immediate heir of ancient Greek mathematics was medieval Arabic mathematics. But Arabic mathematics also had access to much Indian and some Chinese material. And Indian and Chinese mathematics, though in many ways more sophisticated than the Egyptian or Babylonian, still does not emphasize rigorous proof. Unsurprisingly, therefore, Arabic mathematics, even more so than Greek, shows a contrast between a rigorous geometric side and a non-rigorous algebraic side.

On the algebraic side, we find a figure like al-Khwarizmi (eighth–ninth centuries), whose work on Indian numerals, in Latin translation, introduced "Hindu-Arabic" or decimal numerals to Europe, and whose name gives us our word "algorithm." On the geometric side we find a figure like Omar Khayyam (eleventh–twelfth centuries), author of the famous quatrains, but also perhaps the greatest medieval mathematician, and author of a commentary on Euclid and of geometrical solutions to cubic equations.[13]

As for "modern" in the sense of post-medieval mathematics, the tradition that begins with Renaissance Italian mathematics and gradually grows into present-day cosmopolitan mathematics, it inherited geometry in the form of an organized body of beautiful theorems, and algebra in the form of a disorderly mass of useful techniques. The different etymologies, Greek for "geometry" and Arabic for "algebra," partly reflect this dualism, as does the fact that as late as the present author's school days (and, for all I know, even today) the tradition in secondary-school pedagogy was that "geometry has proofs but algebra does not."

The first original, creative modern developments, the solution of cubic and quartic equations by Gerolamo Cardano and others, were on the algebraic rather than the geometric side, and the novel methods introduced, especially the use of "imaginary" numbers, were spectacularly non-rigorous.[14] The next great achievement (apart from the development of modern algebraic notation) was the introduction of coordinate methods by René Descartes and Pierre de Fermat.[15] This permitted the application of powerful algebraic techniques to geometrical problems, though so long as algebra remained non-rigorous, such applications of it inevitably compromised the rigor of geometry. Thus *non-rigorous* early modern mathematics is virtually coextensive with *interesting* early modern mathematics. Non-rigorous early modern mathematics, however, in contrast with the work of Ahmes or Heron, was characterized less by mathematics without proof than by proof without rigor.

---

[13] It should be noted that "Arabic," and even to some extent "Greek," in the present connection refers primarily to the language in which mathematics was written. Ethnically, Omar Khayyam was Persian, and Heron may have been Egyptian. If one applied the same convention to naming sub-periods of modern mathematics, the period down to about 1800 would be called "Latin," the nineteenth and early twentieth centuries "Franco-Germanic," and the period since 1945 "English."

[14] Kline tells the story, in which mathematics mixes with intrigue, in his ch. 13, sect. 4, using modern notation. Coverage of algebra in Smith 1929 begins (201–12) with three short extracts from Cardano; footnotes to the first of these exhibit and explain Cardano's own notation.

[15] Newman 1956c gives excerpts from Descartes (i. 239–53), while Smith 1929 gives excerpts from both Descartes (397–402) and Fermat (389–96). But in the future I will for the most part not note such facts, leaving the reader to explore these sources without further specific references on my part. Something will be found from almost every figure I mention.

Infinitistic Methods

Especially after the development, a generation or so after coordinate methods, of the differential and integral calculus, one found methods or "theorems" supported by considerations or "proofs" that were not rigorous deductions, but involved steps that from a present-day point of view would be considered merely heuristic. When we reach this period of explosive growth, it is very easy to become lost among the trees, while what we need for our ultimate purpose of clarifying the nature of rigor is, rather, a map of the woods, an overview from a rather high altitude, even if such a view inevitably blurs details and sometimes requires simplistic summaries of complex situations. Thus simplifying, we may concentrate to begin with just on the calculus, and just on two main forms in which non-rigorous steps appeared in early modern "proofs" in that subject: the *use of infinitistic methods*, involving systematic employment of what amount to arguments by analogy, and the *appeal to spatiotemporal intuition*.

The process of the subsequent rigorization of the calculus was a long one. It had already begun in the 1780s, when Joseph Louis Lagrange persuaded the Berlin Academy to offer a prize for work on the foundations of the calculus (for which the submissions obtained turned out to be rather disappointing). It was more seriously and systematically pursued by Augustin Cauchy from the 1820s on. But the most influential work was completed only decades later by Karl Weierstrass. Even the latter's work left something more to be done, mainly by Richard Dedekind. It was not until the twentieth century that rigorized calculus began finding its way into undergraduate textbooks, Hardy 1908 being a notable landmark. There will be space here only to note briefly, in connection with each of the two main forms of non-rigor just identified, the approach that was ultimately taken to its elimination over the course of the extended rigorization process.

To begin with another simplification, though the calculus is today attributed to Isaac Newton and G. W. Leibniz about equally, I will simply neglect Newton and his "prime and ultimate ratios" and consider only Leibniz and his "infinitesimals." An excuse for so doing is that the ferocious eighteenth-century priority dispute led to British mathematicians patriotically adhering to the Newtonian notation of "pricked letters," while Continental mathematicians adopted the Leibnizian "d-ist"

symbolism, the notation still in use today. The latter is the more convenient and useful—it gives us more of a "calculus" in the sense of a method
by which problems can be settled by simple calculation, letting the symbols do one's thinking for one—and for this, among other reasons, British
mathematics soon fell behind Continental, and in a short survey may be
left out of the picture.[16]

The operations of the calculus were important to the new physics in
the first instance because differentiation is what we need to get from
a formula expressing some variable physical quantity as a function of
time to a formula for the rate of change of that quantity at any given time
(for instance, from position to velocity or velocity to acceleration); and
inversely, integration is what we need to get from a formula expressing
the rate of change of some variable physical quantity as a function of
time, and the value of that quantity at some one time, to a formula for
the value of that quantity at any other time (for instance, from acceleration and initial velocity to velocity, or from velocity and initial position
to position). But many of the earliest problems treated by differential
and integral calculus were geometric ones, resembling and subsum-
ing problems treated already in ancient Greek mathematics. Typical
examples are the problem, solved by Apollonius, of determining the
tangent to a parabola at a given point, and the question, answered by
Archimedes, of determining the area of the region bounded by a parabola and a line.

The ancients had their own, rigorous methods for dealing with such
problems. Notably Archimedes used what is called the *method of exhaustion* to prove results about areas and analogously about volumes. To
describe his procedure in slightly modernized form, he starts from the fact
that we know how to calculate the areas of polygons. One can then show
that a given region has an area no less than (respectively, no greater than)
a certain value v by showing that for any value slightly less (respectively,
slightly greater) than v, call it $v - \varepsilon$ (respectively, $v + \varepsilon$), where ε is some
small positive number, if we allow polygons with a large enough number
N of sides, then a polygon can be inscribed in (respectively, circumscribed
about) the given region whose area is greater than (respectively, less than)
the given value $v - \varepsilon$ (respectively, $v + \varepsilon$).

---

[16] I will also leave out of account the "method of indivisibles" used by some
pre-Newtonian, pre-Leibnizian mathematicians such as Cavalieri.

The moderns, in attacking tangency and area/volume problems, used coordinate methods. For the tangent problem, the parabola would be represented as the graph of some quadratic equation, perhaps $y = 2 - x^2$. A line is always the graph of some linear equation $y = bx + c$, where b is the *slope* of the line, and where c can be easily determined once b and a single point on the line are known. The problem of determining the tangent reduces to the problem of determining the slope of the tangent at a given point $(x, 2 - x^2)$ on the parabola.

For the latter problem, infinitesimals were introduced, and with curves being treated by analogy to polygons. The parabola was viewed as if it were a polygon with infinitely many infinitesimally small sides, the side at x running from the point $(x, 2 - x^2)$ to the point $(x + d, 2 - (x + d)^2)$. The tangent line at x will simply be the line extending this side in both directions. The slope is easily calculated, thus:

$$((2 - (x + d)^2) - (2 - x^2))/(x + d) - x) =$$

$$(x^2 - (x + d)^2)/d =$$

$$(x^2 - (x^2 + 2dx + d^2))/d =$$

$$-(2dx + d^2)/d =$$

$$-2x - d$$

Neglecting the infinitely small term d, the final answer is given as $-2x$. Similarly, the problem of determining the area of the region bounded by the curve $y = 2 - x^2$ and the x-axis or line $y = 0$, is approached by thinking of the region as made up of a row of infinitely many infinitesimally narrow rectangles, side by side, with the height of the rectangle at x being $2 - x^2$.

Such methods conspicuously lack rigorous justification. The calculation of the slope of the tangent seems egregiously illogical, because d is first implicitly assumed to be nonzero in the series of equations just displayed, since we divide by d while one cannot divide by zero, and then immediately afterwards d seems to be implicitly assumed to be zero, since it is neglected in giving the final answer. This is the sort of procedure that famously became target of criticism of George Berkeley in his *Analyst* (though naturally Berkeley, unlike us, is especially concerned with *British* mathematics).

On Berkeley's sort of account, the answer to the tangent problem that we obtain is correct, but only because of compensation of errors. On the one hand, a parabola is *not* a polygon. On the other hand, $2x + d$ is *not* equal to $2x$. And yet the slope of the tangent *is* $2x$. "By a twofold mistake he [the mathematician] arrives, though not at Science, yet at Truth." Though the purpose of the *Analyst* is perverse—Dean Berkeley aims to justify a lack of rigor in theology by pointing to the lack of rigor in mathematics—the critical observations are acute. If they had little influence, it was not because they lacked cogency, but more likely because they were unaccompanied by the offering of any viable systematic alternative to the methods criticized.[17]

What amounts to an acknowledgment of lack of rigor in the calculus, accompanied by a claim to the effect that the missing rigor could be supplied if one thought it worthwhile to take the time to do so, can be found in Leibniz himself, in a communication much quoted by commentators:

for instead of the infinite or the infinitely small one takes quantities as large and as small as needed to make the error less than the given error, in such a way that one differs from the style of Archimedes only in the expressions, which are more direct on our method and more suited to the art of discovery.[18]

Talk of infinitesimals is here in effect excused by claiming it to be no more than a *façon de parler*, a kind of shorthand for something more rigorous. (Leibniz generally took the notion of infinitesimal less literally than did many of his disciples.)

In the style of Archimedes, described in what is admittedly a somewhat modernized form, the "given error" would be the ε, and the magnitude that has to be "so large" as to make the error that small would be the N. Leibniz does not actually demonstrate how one is supposed to be able to translate from his style to that of Archimedes, and indeed, the calculus of Leibniz and other "moderns" treats so many more kinds of problems than

[17] Extracts from the *Analyst* are given both in Newman 1956c (i. 288–94) and Smith 1929 (627–34). Jesseph 1993, among other contributions, gives a hint of what a mathematics founded on Berkeley's idea of the *minimum visibile* might have been like.

[18] *Car au lieu de l'infini ou l'infiniment petit, on prend des quantités aussi grandes et aussi petites qu'il faut pour que l'erreur soit moindre que l'erreur donnée, de sorte qu'on ne diffère du stile d'Archimède que dans les expressions, qui sont plus directes dans notre méthode et plus conformes à l'art d'inventer* (Leibniz 1701). The passage is discussed in Robinson 1967, to which I will allude later.

did the ancients that it is in many cases quite unclear what exactly would be *meant* by redoing a problem in "the style of Archimedes."

The last thing Leibniz says, however, about his method being more useful for discovery, is certainly true. To apply the method of exhaustion, one must know the answer one is trying to establish, while the use of integration enables one to calculate that answer. And though Leibniz didn't know it and couldn't have known it, since the key manuscript (Heath 1912) was discovered only in 1905, Archimedes himself used something not unlike infinitesimal considerations for the *discovery* of key results, and only wrote them up in the rigorous, if exhausting, "exhaustion" style afterwards. Seventeenth- and eighteenth-century analysts simply omitted the final step here; but that means that they were, from the standpoint of a rigorist like Archimedes, failing to justify by proof the results they had discovered by a heuristic method.

So much, for the moment, about the role of the infinitely *small*. It remains to say something more about the role of the infinitely *large*, and specifically about the role of infinite *series*. We may begin with those for the trigonometric functions:

$$\sin x = x - x^3/3! + x^5/5! + \cdots$$

$$\cos x = 1 - x^2/2! + x^4/4! + \cdots$$

Taking any finite number of terms on the righthand side, we obtain a polynomial function that approximates the sine or cosine function (over a part of its domain), as the case may be. The more terms one takes, the higher the degree of the polynomial one considers, the better the approximation. It is a short step—but a decisive one, since it takes us from rigor to non-rigor—to treating the trigonometric functions by analogy with polynomials, as if they were polynomials of *infinite* degree that could be manipulated just as one would other polynomials.

For instance, quite early it was worked out how to differentiate or integrate a monomial term ax^n, and was proved that for sums of *finitely* many functions, the derivative or integral of the sum will be the sum of the derivatives or integrals of the given functions. Hence polynomials of *finite* degree could be differentiated or integrated "term-by-term." This procedure was then extended to "polynomials" of infinite degree, and used to conclude, for instance, that the derivative of the sine is the cosine (since the derivative of x is 1, the derivative of $x^3/3!$ is $x^2/2!$, and so on) and similarly that the derivative of the exponential function is just itself. And though those particular results can be obtained in other ways, "termwise" differentiation and integration were more essential in other cases. Moreover, infinite series were manipulated like finite polynomials not only in differentiation and integration problems, but in many other ways.

The paradigm, or stock example of a successful use of such manipulations, is the derivation by Leonhard Euler of a series for π:

$$1/1^2 + 1/2^2 + 1/3^2 + 1/4^2 + \cdots = \pi^{2/6}$$

The tale of this example has been told well by Polya (1954, i, sect. 2.6) and I do not need to refer to any details of it here. The example provides an instance where considerations not at all amounting to a rigorous proof are nonetheless very convincing indeed. Another commentary, Steiner 1975 (ch. 3, sect. 4), goes so far as to say that Euler's heuristic argument produces not merely *confidence* but *knowledge*.

The foil, the stock example of *un*successful use of similar manipulations, is provided by the alternating series

$$x = 1 - 1 + 1 - 1 + 1 - 1 + \cdots$$

Here we evidently have

$$
\begin{aligned}
x &= (1-1) + (1-1) + (1-1) + \cdots \qquad &\text{(A)} \\
&= 0 + 0 + 0 + \cdots \\
&= 0
\end{aligned}
$$

But note also that we also have

$$
\begin{aligned}
x &= 1 - 1 - (-1) - 1 - (-1) - 1 - (-1) \cdots \qquad &\text{(B)} \\
&= 1 - (1 - 1 + 1 - 1 + 1 - 1 + \cdots) \\
&= 1 - x
\end{aligned}
$$

And $x = 1 - x$ implies $2x = 1$ and $x = \frac{1}{2}$. A genius of the caliber of Euler could obtain correct results by judicious use of non-rigorous infinitistic methods that in other hands could be used to "prove" that $0 = \frac{1}{2}$ or worse.[19]

In the subsequent nineteenth-century rigorization of analysis, infinitesimals were completely eliminated in favor of various versions of the policy

---

[19] Geniuses of later times have even found a use for the absurd calculations in (A) and (B), in what is called the "Eilenberg-Mazur swindle" in topology, but that matter is beyond the scope of the present discussion. A sense, not dependent on the calculation (B), in which the series can be considered "summable" to $\frac{1}{2}$ will be explained in Ch. 2.

of taking "instead of the infinite or infinitely small quantities as large or as small as needed to make the error less than the given error." In the rigorization of series manipulations, by contrast, the word "infinity" and the symbol "∞" were not eliminated, but they came to be taken non-literally. Just what I mean by saying this will require a bit of explanation.

In the treatment of series, the first step towards rigorization was to define rigorously a distinction between good series (like the Euler series for π) and bad series (like the alternating series). This rigorous treatment of infinite series presupposed a rigorous treatment of infinite sequences. The relevant mathematical material can be found in almost any calculus textbook, and I will say only as much about its content as is needed to bring out certain philosophical points. Let us begin with the example of the following sequence (given first in common fractions, then in decimal notation):

$$2/3, \quad 3/5, \quad 4/7, \quad 5/9 \quad \ldots$$
$$.666\ldots, \quad .600\ldots, \quad .571\ldots \quad .555\ldots \quad \ldots$$

Here the general term is $a_n = (n + 1)/(2n + 1)$. Even after rigorization, one still says "the limit of the sequence as n goes to infinity is one half" and one still writes

$$\lim_{n \to \infty} a_n = 1/2$$

and so on. But this is explained to mean that for any given error $\varepsilon > 0$ there is a number N so large that for all $n \geq N$, the term a_n will be within ε of one-half, so that $|a_n - \frac{1}{2}| \leq \varepsilon$. For instance, if we want the terms to be within one decimal place of the limit ($\varepsilon = 10^{-1}$), it is enough to go out to the third term ($N = 3$); after that, they all have a 5 in the first decimal place. Note that the word "infinity" and the symbol "∞." though parts of the expressions to be explained, drop out and do not appear in the explanation. This is what was meant by saying that these expressions are taken "non-literally"; a fancier label is that they are "contextually defined."

A sequence like the one given, which goes to some limit as n goes to infinity, is called convergent. A sequence like 1, 2, 3, 4, ... or 1, 0, 1, 0, ... is called non-convergent (divergent or oscillating). This distinction for sequences is then applied to series. The series

$$a_1 + a_2 + a_3 + a_4 + \cdots$$

is said to sum or converge to b if the sequence of *partial sums*

$$s_1 = a_1$$
$$s_2 = a_1 + a_2$$
$$s_3 = a_1 + a_2 + a_3$$

goes to b as n goes to infinity. Subtler distinctions can be made, between what are called "conditional" and "absolute" convergence, and the range of series to which various manipulations can be applied compatibly with the requirements of rigor is then identified in terms of convergence conditions.

The fact that one treats all mentions of "infinity" as non-literal was sometimes called the rejection of *actual* infinity, while the fact that one still works with a sequence for which there is no bound to the number of terms was by contrast called accepting *potential* infinity. So we may say that rigorization involved the banning—from *proofs*, for they never disappeared from *heuristics*—of actual infinity as well as of infinitesimals.

But all this makes it seem as if what was wrong, from the point of view of rigor, with infinitistic methods was the positing of some kind of dubious *entity*, infinitely small or infinitely large. Such a conclusion would confirm the prejudice of many philosophers that overwhelming importance attaches to issues of what they are pleased to call "ontology," assumptions about the existence of entities of one sort or another. To see the problem as ontological would, however, be to misperceive it.

That a misperception would be involved is clear from the fact that in the twentieth century the logician Abraham Robinson showed (1966) how it is possible to achieve complete rigor in the calculus even while *retaining* infinitesimal and actually infinite quantities. He developed a whole "nonstandard analysis" based on such principles, and H. J. Keisler even wrote a rigorous freshman calculus textbook based on rigorized, Robinsonian infinitistic methods (see Keiser 2000).

Though Robinson was a logician, Robinson's rigorization does not involve adopting some kind of deviant logic, a "paraconsistent logic" in which d can be both equal and not equal to zero, or in which x can be equal both to 0 and to ½. Rather, it consists in making certain distinctions. One must be careful to distinguish, in the case of two quantities x and y, their being strictly equal, $x = y$ from their being infinitely close, $x \approx y$. Above all, one must be careful about which laws applicable to finite quantities can be extended to the infinite and infinitesimal. What modern—now called "classical"—logic supplied Robinson was an apparatus sufficient to characterize rigorously the class of laws that do carry over: They are the ones expressible in a certain logical notation and symbolism.

What was wrong with infinitistic methods was obscurity about just what the postulates pertaining to such entities were supposed to be, confusion about what one was supposed to be allowed to do with the entities posited, and not the positing of the entities as such. Robinson in effect did for Leibniz's heuristic principles for the manipulation of infinitesimals what Weierstrass did for Leibniz's quoted remarks about eliminating infinitesimals in favor of ancient methods of exhaustion.[20] The two approaches, with or without the infinitesimals, are equally capable of being made fully rigorous.

Infinitistic methods, treating infinite series like finite polynomials, and curves and the areas under them like broken lines and polygonal figures, are evidently special kinds of argument by analogy, one of the forms of non-rigorous reasoning emphasized by Polya. The use of such methods, and other aspects of early modern mathematics, also exemplify what I will call reliance on *generic reasoning*. By this I mean performing calculations or manipulations that usually or generally work but sometimes or exceptionally fail, treating as universally applicable techniques that are—and may be *known* to be—only applicable in favorable cases, in the hope and belief that the case in which one is interested is one of the favorable ones, but without any rigorous justification for supposing this to be so. And what rigorization required was making certain clear distinctions, identifying precisely and exactly the scope of the methods of calculation or manipulation in question, and the limits beyond which they cannot be safely applied. That is what Weierstrass and Robinson alike, each in his own way, was doing when he succeeded in developing the calculus on a rigorous basis.

As will be seen later, what I am calling generic reasoning, the sloppy use of methods that were at best typically or normally applicable, as if they

---

[20] Just how far Leibniz may be regarded as anticipating Robinson is a matter of some dispute among historical scholars. Robinson 1967 is quite generous in this regard. Leibniz did make something like the distinction between = and ≈. He certainly did not give a *rigorous* characterization, comparable to that of Robinson, of the class of laws that carry over.

were unqualifiedly and universally applicable, was engaged not only in connection with the calculus, but also in other areas of analysis beyond the calculus, and in areas of mathematics beyond analysis. It was a major form, and perhaps *the* major form of non-rigor in early modern mathematics, more significant perhaps than reliance on nonverbal thought, sudden insight, inductive generalization, or the like. The only form of non-rigor to rival it in importance was reliance on *intuition*, to which let us now turn.

Spatiotemporal Intuition

In approaching the topic of intuition we must be careful about terminology. In ordinary English, 'intuition" covers just about any form of purported knowledge of whose source we are not consciously aware and can give no account, any hunch or instinctive belief or gut feeling not consciously inferred from anything more basic. But the word has another use in philosophical English, as the translation of the word *Anschauung* as used in philosophical German, and especially as used by Immanuel Kant and in the Kantian tradition. This use of "intuition" is more closely related to the Latin etymology of the word (from *intueri, to look*) than to its meaning in present-day colloquial English. When used in this Kantian sense, the term means something like (the faculty of) "sense-perception," and is associated with awareness of the general form of perceptual space and time.

One key difference between the two meanings is that a hunch is generally personal to one subject, while the faculty of spatiotemporal perception is supposed to be shared by all (human) subjects. To allow in one's arguments steps of the form, "At this point I personally have a hunch that . . . " would be to give up on the idea of "proof" in any sense implying an interpersonally valid justification, one that can be shared with others. To appeal to our presumably shared human faculty of spatiotemporal intuition would not require abandoning "proof" in this minimal sense, though it certainly would involve abandoning rigor. It is this latter kind of "intuition," *spatiotemporal* intuition, that is pertinent here.

Infinitistic methods generally, and the use of infinitesimals in particular, were not intuitive in *either* sense of the word, and can be called "intuitive" only if one uses that term so broadly and loosely as to make it nothing more than an unneeded synonym for "heuristic." Infinitistic considerations always involved some form of argumentation or calculation, and never an *immediate* cognition or pretended cognition, not derived from anything more basic, and so were not "intuitive" in the first sense noted here. Neither were they "intuitive" in the Kantian sense: A parabola does not *look like* a many-sided polygon, nor does the area bounded by a parabola and a line *look like* a row of narrow rectangles. The curve looks smooth and not jagged. And there is nothing especially spatiotemporal about infinite series manipulations.

Yet appeal to spatiotemporal intuition, though not involved in the specific matters I have discussed to this point, was very much present in analysis before the nineteenth-century campaign for rigorization. Actually, we need to make a threefold distinction here between (1) *temporal* or *chronometric* intuition; (2) *spatial* or *geometric* intuition; and (3) *spatial-plus-temporal* or *kinematic* intuition. Chronometric intuition played no large role in eighteenth- or nineteenth-century developments, though it did in the philosophy of Kant; in the work of William Rowan Hamilton, the mathematician after whom Hamiltonians in mathematical physics are named (Hamilton 1837); and later in the work of L. E. J. Brouwer, whom we will meet in Chapter 2. The main distinction to be drawn is between the roles of geometric intuition in connection with the conception of the real numbers as represented by points on a line, and of kinematic intuition in connection with the conception of functions as represented by curves in the plane generated by a moving point.

The notion of a *continuous* function was generally defined in non-rigorous infinitesimal analysis as a function f such that the absolute difference $|f(a) - f(x)|$ between its values for two arguments is always infinitesimal whenever the absolute difference $|a - x|$ of the arguments themselves is infinitesimal. Something not unlike this definition applies today in rigorous nonstandard analysis as well (but with a careful distinction being made between strict equality and infinite closeness). Intuitively, however, continuity was understood kinematically: A continuous function is one whose graph can be generated by a point moving in the plane *without ever jumping*. A simple example of *dis*continuity in this sense is provided by the function $y = [x]$ that for argument x takes as value the *greatest integer* less than or equal to x. This function "jumps" at every integer, its value remaining 0 for instance, for $x = 0.9$ or 0.99 or 0.999, but suddenly becoming 1 for $x = 1$.

The notion of a *differentiable* function also had a definition in terms of infinitesimals. If $y = f(x)$, then the derivative $y' = f'(x)$ was taken to be dy/dx, where dx is infinitesimal and dy is the infinitesimal difference $f(x + dx) - f(x)$. Again, something not unlike this definition applies today in rigorous nonstandard analysis, and the notation dy/dx as a whole symbol-complex may be used for the derivative even in rigorous standard calculus or analysis down to this day, though the derivative is no longer thought of literally as a ratio of infinitesimals. But intuitively, differentiability was understood kinematically: The moving point must not only avoid jumping, but must also have a definite direction of motion at each point, and avoid stopping and turning abruptly. A simple example of *non*-differentiability in this sense is provided by the *absolute value* function $y = |x|$, which abruptly changes direction at the origin point where $x = y = 0$: To the left, for negative x, it slopes *down*wards at a 45° angle, while to the right, for positive x, it slopes *up*wards at a 45° angle. There is no well-defined direction of motion at the origin itself. In geometric terms, there is no single tangent line, no single line that touches the graph without intersecting it nearby, but rather a whole range of such lines.

Note that, as the simple examples cited show, continuity and differentiability are properties that a function can have at some points and lack at others. What kinematic intuition seemed to reveal to many mathematicians was that a curve that never jumps can only stop and make an abrupt change of direction a *finite* number of times in any given interval: A function continuous at every point in an interval can only be non-differentiable at finitely many points in that interval. André-Marie Ampère, after whom the electrical unit the amp is named, even announced a "theorem" to this effect. The "proof," however, cannot have been rigorous, for the "theorem" is false.

This was shown by a counterexample of Bernhard Riemann, improved by Weierstrass to an example where a function, though continuous everywhere, is differentiable nowhere.[21] Not unrelated is another famous example, Giuseppe Peano's *space-filling curve*. Defying intuition, it manages to

---

[21] Work of Bernhard Bolzano, some of it published in his lifetime, most of it not, contains anticipations, often decades ahead of anyone else, of various developments in the nineteenth-cent. process of rigorization, including examples of continuous non-differentiable functions antedating Riemann's. But most nineteenth-century mathematicians seem to have been unaware of Bolzano's work, or indeed of his very existence, and since he thus played little direct role in the processes I am describing, I leave him out of my abbreviated account.

visit every point in the interior of a square in a finite amount of time, and without any jumps. Examples like these are prominent in Hans Hahn's popular essay "The Crisis in Intuition" (1956a).

The examples given by Hahn are, however, visually *illustrated* in his article by drawings, not indeed of the counterexample functions themselves, but of successive approximations to them. It might therefore be said by a champion of intuition that Ampère's "theorem" only *seemed* to be dictated by intuition, whereas a more refined employment of intuition—pondering the drawings in Hahn's article, for instance—reveals that the "theorem" fails. It might even be suggested that Riemann and Weierstrass and Peano themselves probably employed their refined spatiotemporal intuitions in discovering their counterexamples. And all this may very well be true.[22]

But it is largely irrelevant for present purposes. For all one ever has to go on, if one appeals to intuition, is one's *apparent* intuitions at the time. If "apparent" intuitions are not all "real" intuitions, then one is going to need something *other* than intuition—something like rigorous proof—to sort out cases and distinguish which apparent intuitions are real ones.

Weierstrass replaced the old definition of continuity at an argument a

> $f(x)$ will be only infinitesimally different from $f(a)$ whenever
> x is only infinitesimally different from a

by a new one

> $f(x)$ can be kept as close as desired to $f(a)$
> by taking x close enough to a

and, more importantly, explained the latter formulation more formally as meaning

> for every $\varepsilon > 0$ however small ("as close as desired")
> there is a $\delta > 0$ ("close enough") such that
> $f(x)$ will be within ε of $f(a)$ whenever x is within δ of a

and this amounts to another manifestation of the idea of taking, instead of the infinitely small, a finite quantity (δ) so small as to make the error less

---

[22] Something like the line of objection sketched here appears in developed form in Mandelbrot 1982.

than the given error (ε). There are similar "ε-δ-definitions" for differentiability and other notions.

With the aid of such definitions, Weierstrass was able to make many crucial distinctions. For instance, a function is *pointwise* continuous on an interval if for every x in the interval and for every $\varepsilon > 0$, however small, there is a $\delta > 0$ such that (and so on as shown). By contrast, a function is *uniformly* continuous on an interval if for every $\varepsilon > 0$, however, small there is a $\delta > 0$ such that for every x in the interval (and so on as before). The distinction between pointwise and uniform continuity was one with which Weierstrass's predecessors in the rigorization movement, such as Cauchy, had struggled in vain. The distinction is a matter of the order of quantifiers (all-all-exists *vs* all-exists-all, or in logicians' notation $\forall\forall\exists$ *vs* $\forall\exists\forall$) and cannot be made unless the quantifiers have been made explicit and not left implicit in expressions like "close as desired" or "close enough." With the aid of such distinctions, Weierstrass was able to state precise conditions for the validity of various results that previously had only been thought of as holding generically but not universally, cleaning up much sloppiness that had previously affected analysis.

I earlier alluded to the fact that many students of mathematics today get their first introduction to rigor in a course on analysis, which is to say, a course where the calculus is done or redone minus appeals to intuition, entirely in terms of epsilons and deltas. Nonetheless, ε-δ-definitions do not by themselves entirely eliminate the need for appeal to intuition: They may eliminate the need for appeal to *kinematic* intuition, and they may reduce the need for appeal to *geometric* intuition to a single principle, but further work would be needed to get the principle in question entirely without appeal to intuition. Let me elaborate.

A crucial basic result in analysis is the *intermediate value theorem*, according to which if a continuous function f takes a negative value $f(a) < 0$ for some real number a and a positive value $f(c) > 0$ for some real number c with $a < c$, then it must take the zero value $f(b) = 0$ at some b between them, $a < b < c$. The result seems almost immediate if appeal to kinematic intuition is allowed. For then it just says that if a point moves from a position below the x-axis to a position above the x-axis, and does so without ever jumping, then it must sometime *meet and cross* the x-axis. How else can it get from one side to the other without meeting the axis, if it does not jump over it?

What the ε-δ-definition of continuity accomplishes is to reduce the task of proving this theorem rigorously to the task of proving another theorem rigorously, the *completeness principle*, as it is sometimes called. According this result, if we make a *cut* in the real numbers, which is to say, if we divide the real numbers into two classes, a *lower* and an *upper* class, in such a way that every member of the lower class is smaller than every member of the upper class, and every member of the upper class larger than every member of the smaller class, then there must be a *boundary* number, either a largest number in the lower class, or a smallest number in the upper class.

The nature of the reduction of the intermediate value theorem to the completeness principle is perhaps sufficiently illustrated by the special case where the function f is *monotone*, so that we always have $f(x) < f(y)$ whenever $x < y$. In that case, if we had a failure of the intermediate value theorem, if there were no x with $f(x) = 0$, then the division between the x with $f(x) < 0$ and the x with $f(x) > 0$ would by monotonicity be a cut: Every number for which the value of f is negative would be less than every number for which the value of f is positive, and every number for which the value of f is positive would be greater than every number for which the value of f is negative; moreover every number would be of one kind or the other. Continuity of f, once rigorously defined, can be rigorously shown to imply that the lower class has no largest member, that given any number for which the value of f is negative there is a slightly larger number for which this is still the case, and that the upper class has no smallest member, that given any number for which the value of f is positive there is a slightly smaller number for which this is still the case. But that means that we have no boundary point between the two classes, and a violation of the completeness principle. Thus if the completeness principle cannot be violated, neither can the intermediate value theorem.

As for the status of the completeness principle itself, it seems immediate if appeal to spatial intuition is allowed. For in geometric terms, what the principle says is that if we divide a horizontal line into two pieces, with every point in the first lying to the left of every point in the second, and every point in the second lying to the right of every point in the first, then there must either be a rightmost point in the first, left class, or a leftmost point in the second, right class. And this seems intuitively evident. But to eliminate this one last appeal to intuition, completing the "degeometrization" or "arithmetization" of analysis, more work was needed by Weierstrass's successors, notably Dedekind, to whom we will return later.

Ancient Quasi-rigor

To sum up the discussion to this point, we have so far seen a rough initial characterization of rigor as requiring deduction of new results from old, and ultimately from postulates, with an analogous requirement on definition. We have also seen examples of various forms of non-rigor (beyond simple appeal to testimony): nonverbal thought, sudden insight, inductive generalization, analogical extrapolation, what I have called generic reasoning (though it really might just be called "sloppiness"), and lastly spatiotemporal intuition. We have noted that these forms of non-rigor were widespread until quite late in the nineteenth century at least, and that it took quite a bit of work to eliminate dependence on them.

The question *why* mathematicians allowed themselves to use such methods is easily answered by citing the methods' great utility, already mentioned by Leibniz: their ability to provide useful results. It is hard to imagine how seventeenth- or eighteenth-century mathematicians could have met the pressing mathematical needs of the rapidly developing physics of the period if they had insisted pedantically on rigor. The more difficult question would seem to be why mathematicians over the course of the nineteenth century decided to adopt more rigorous methods. We may be glad that they did so, because it is hard to imagine mathematicians developing Riemannian manifolds or Hilbert space, the mathematical tools of general relativity and of quantum mechanics, if they had remained content with the standards of rigor of circa 1800; but needless to say, when the rigorization process was beginning, no one knew these things were coming.

Before going deeply into this question of why nineteenth-century mathematicians went in for rigor, however, we should ask the prior questions (1) why ancient Greek mathematicians adopted the ideal of rigor in the first place; and (2) why later mathematicians continued to honor that ideal in principle, even if they strayed from it in practice.

The answer to question (2) seems to be, disappointingly, no more than "because of the weight of tradition," or "because of the authority of the ancients." And we may point to a single work, Euclid's *Elements* (see Heath 1926) as the chief concrete embodiment of the tradition and authority. Throughout the early modern period and on into the nineteenth century, Euclid's *magnum opus* was admired for its aesthetic charm even by political theorists such as Thomas Hobbes, and endorsed as providing excellent training for the mind even by statesmen such as Abraham Lincoln. It was the most successful textbook in history, subject to endless editions and translations and commentaries.

The answer to question (1) may be, even more disappointingly, that we will never know. Almost all that we have left from the earliest stages of Greek mathematics consists, not in copies of works from that period, but only in passages *about* such works coming from later (sometimes very much later) antiquity, passages called "fragments" or "testimonies" according as they do or do not purport to be direct quotations from works now lost. The unfortunate fact is that the fragments and testimonies available for early Greek mathematics (as in Thomas 1939) may simply be insufficient to support a definite conclusion, either about when the ideal of rigor was adopted or why.

If we ask *why* we are so lacking in early sources, Euclid's *Elements* must be mentioned again. For the most probable answer is that the success of that work led to neglect of its predecessors and rivals, which eventually ceased to be recopied. We do at least have reliable evidence that earlier and presumably less successful books of *Elements* existed before Euclid's, and in particular that there was one by Eudoxus, an associate of Plato to whom the theory of proportion found in book 5 of Euclid is attributed. Because Eudoxus comes a generation before Aristotle, whereas Euclid comes a generation after, if we had the older *Elements* scholars might be able to judge how far Aristotle in the *Posterior Analytics* is giving an idealized account of contemporary mathematical practice, and how far he is prescribing the way things ought to be done rather than describing the way things are done. Scholars can make this kind of comparison between the account of tragedy in Aristotle's *Poetics* and the surviving plays of Euripides, for instance.

Also, if we had the older *Elements* scholars might be able to judge how far the similarities between what Euclid does and what Aristotle says should be done are the result of Euclid following Aristotle. It is only a slight exaggeration to say that no modern mathematician would ever follow a philosopher, setting aside the cases of Descartes and Leibniz, where mathematician and philosopher were united in one person; but things may have been different in antiquity, so the suggestion that Euclid may have been following Aristotle is not absurd. But neither is it obviously correct, since the similarities between Aristotle and Euclid may result from their both following Eudoxus or the Greek geometric tradition at large.

Euclid himself just does what he does, without saying why he is doing it. Presumably in his day it was already the expected thing to do.

It is generally supposed that Eudoxus employed the same definition–theorem–proof format as Euclid, and indeed this is presumed not only for the lost *Elements* of Eudoxus but also for the still earlier lost *Elements* of the shadowy Hippocrates of Chios. But whether the work of Hippocrates was the first book of *Elements*, and how much theorem-proving activity there was even earlier, are open and perhaps unanswerable questions. And if the *when* of the introduction of rigor is in doubt, the *why* is much more so.

A "just so story" used to be told about the matter, running roughly as follows. It all goes back Pythagoras and his Pythagorean Fraternity/Sorority. On the one hand, the group had a general program that involved—besides eschewing beans—developing an explanation of the world in arithmetical terms. Everything was to be accounted for in terms of numbers, by which were meant whole numbers, *one, two, three,* and so on, and in terms of ratios of numbers, such as the 1:2 and 2:3 and 3:4 ratios corresponding to the musical intervals of the octave and fifth and fourth. The arithmetical program was supposed to apply to geometric matters as well, witness discussions of triangular and square and pentagonal numbers, for instance.

A ratio of two line segments AB and AC could be equated with a ratio of numbers provided a *common measure* could be found, a segment AD such that some whole number m of copies of it would exactly fill up AB while some (other) whole number n of copies of it would exactly fill up AC. For the ratio of AB to AC would then be the ratio of m to n.

On the other hand, the founder had brought back from his travels knowledge of the discovery that the square on the hypotenuse of a right triangle is equal in area to the sum of the squares on the other two sides. A corollary was noticed, that the big square on the diagonal of a little square would be exactly twice as large. And then a corollary of the corollary was obtained, that the diagonal and the side of a square are *incommensurable*, having no common measure. And thus the group's program was disrupted. Sacrifices of oxen, vows of secrecy, and shipwrecks followed. So also followed the two most peculiar features of Greek mathematics: Geometry came to be recognized as more general than arithmetic (so that Euclid in his books 7 through 9 presents even number theory with geometrical diagrams); and rigorous proof and not just intuitive plausibility came to be insisted upon.

How much of this is true? A century or so ago, scholars could tell us a great deal about Pythagoras and his views on mathematics and cosmology. The mathematician, logician, and metaphysician Alfred North Whitehead, in particular, was enthusiastic about the ancient sage, declaring:

He insisted on the importance of the utmost generality in reasoning, and he divined the importance of number as an aid to the construction of any representation of the conditions involved in the order of nature. (Whitehead 1925, 41)

Such has been the progress of classical scholarship in the intervening years that today what we can say with confidence about Pythagoras as mathematician or cosmologist is precisely nothing.

Whitehead did not just invent his picture of Pythagoras, but based it on sources from later antiquity. Subsequent scholarship, however, tracing back behind those sources to *their* sources, has found the trail to lead back to members of Plato's Academy, generations after the lifetime of the legendary sage, but no further.[23] There were indeed some mathematicians, notably Archytas of Tarentum, active about a century and a half after the time of Pythagoras, who themselves claimed to be or were by others claimed to be "Pythagoreans," but it is not very clear what such a label was supposed to imply, and how much of what it was intended to imply is true.

The question why rigor was first insisted upon is thus one to which we may never be able to give confident answers, barring the discovery of a cache of mathematical Dead Sea Scrolls. And we have seen that we may be able to give only simplistic ones to the question why Euclid continued to be much admired, if not much imitated. The story is further complicated by the fact that admiration for Euclid was fading by the time the nineteenth-century campaign for rigorization got under way. For the awful truth, long known to the more astute commentators, is that Euclid, the paragon and paradigm of rigor, cited by all mathematicians who honored the ideal of rigor from Archimedes and Apollonius onward, departed seriously from that ideal by making essential use of appeals to spatiotemporal intuition, especially in the earliest, must fundamental material.

The closer one looks at the text of Euclid, the stranger his standpoint must seem to us. In the standard English translation by Sir Thomas Heath

---

[23] But see Huffman 2014 for a slightly more optimistic assessment, as well as a chronological list of sources.

(1926), his fine-print notes on book 1 take up several times more space than Euclid's text. Scholars have tried to explain such mysteries as the following: Why does Euclid start from unproved propositions but recognize no undefined notions, offering instead some definition for every notion? For he famously opens his work by saying, "A point is that which hath no part," though this definition can play no role and does play no role in the logical development that follows. Why does Euclid state his first proposition among others as a *problem*, "Given any line segment, to construct some equilateral triangle having that segment as one of its sides," rather than an *existence theorem*, "Given any line segment, there exists some equilateral triangle having that segment as one its sides"?

We may leave such questions to the specialists. The feature of the earliest parts of Euclid most relevant for present purposes is the tacit appeal to intuition in the proof of Proposition 1, and the more overt appeal in that of Proposition 4. In proving Proposition 1, constructing an equilateral triangle with line segment AB as one side, Euclid takes the circle with center at A passing through B and the circle with center B passing through A, and shows that if C lies on both circles, then ABC is equilateral. None of his stated assumptions, however, guarantee that there will *be* any point C lying on both circles. The assumption that there will be one is smuggled into the wording of the proof. A fully rigorous modern approach would acknowledge the need for an additional postulate here in order to guarantee the existence of the point of intersection.

In proving Proposition 4, the side-angle-side criterion for congruence of triangles, Euclid starts with triangles ABC and DEF with AB equal to DE and AC equal to DF and angle BAC equal to angle EDF, and then imagines "applying" the latter angle to the former: in other words, picking the latter up and superimposing it on or fitting it to the former, so that D rests atop A, and the line through D and E atop the line through A and B, and the line through D and F atop the line through A and C. He then argues, among other things, that E will be atop B, because of the equality of AB and DE, and similarly F will be atop C, showing side BC equal to side EF. Nothing in any of his stated assumptions justifies any of his business about superposition. A fully rigorous modern approach would acknowledge the proposition as an additional postulate.

In both cases, the logically questionable steps are intuitively correct. It should be noted that these tacit appeals to intuition go contrary not only to modern notions of rigor, but to the prescriptions of Aristotle. Whether

Euclid himself would have regarded them (if the question were raised) as violations of proper mathematical procedure is a question more difficult to answer. The view of commentators seems to tend to be that the reliance on intuition in Proposition 1 may be unconscious, while that in Proposition 4 is made with a bad conscience, so to speak. In other words, Euclid may be unaware or only half-aware that he is assuming something not implied by his stated postulates in the former case, but is aware that he is doing something fishy in the latter case. What suggests this latter conclusion is that, while the method of superposition would rather obviously make the proofs of many other theorems easier, Euclid avoids it when he can and resorts to it only in a very few cases where he cannot avoid it.

The fact that there are appeals to intuition, which is to say lapses in rigor, even in Euclid means that there was effectively *no* perfect rigor in existence at the time the nineteenth-century rigorization campaign began. Euclid came closer to the ideal than anyone, and yet did not fully realize it. This means that there was no exact precedent for doing what the rigorists wanted to and were trying to do. Something like this last observation is what Russell has in mind when he opens his essay "Mathematics and the Metaphysicians" with these words:

The nineteenth century, which prided itself upon the invention of steam and evolution, might have derived more legitimate title to fame from the discovery of pure mathematics. (Russell 1956, 1576)

The lack of precedent makes the eventual successful completion of the campaign of rigorization a greater achievement.

Belated Rigor

Not only was the degree of rigor sought unprecedented, but the scope of the rigorization project was vastly larger than has so far been indicated. For while the rigorization of the calculus was essentially complete by the last decade of the nineteenth century, there remained other branches of mathematics, including some subbranches of analysis beyond the calculus, to be rigorized, and in some cases their rigorization required quite a bit of additional work and came only considerably later.

A famous case concerns maximization and minimization problems and the use in this connection of the so-called *Dirichlet principle*, a non-rigorous method that came to be increasingly relied on over the course of the nineteenth century, even as the core of mathematics was being rigorized. The principle was invoked by mathematicians of the caliber of Riemann (in the proof of the important *Riemann mapping theorem*), and though a counterexample was found by Weierstrass, it continued to be used sloppily even after it was recognized to hold only generically and not universally. It was only put on a rigorous basis by Hilbert around 1900.[24]

Another famous case is that of probability theory, which prior to the twentieth century was often thought of less as a branch of mathematics than as a separate science that made heavy use of mathematical methods. It took until the middle of the century for it to be made rigorous through incorporation into mathematical analysis, which had itself by this time been made rigorous. The incorporation was accomplished by Andrey Kolmogorov, whose work was first published in the 1930s (originally in Russian) but did not become universally accessible or accepted until the 1950s.

An especially notorious case of tardy rigorization is that of *algebraic geometry*, where the level of rigor in the work of the Italian school, the most active contributors to the field in the late nineteenth and early twentieth centuries, declined as leadership passed from Guido Castelnuovo to Federigo Enriques to Francesco Severi. Things finally reached the point of the last-named in the 1930s and 1940s publishing "theorems" that are demonstrably false, and to which others (David Mumford, Wolf Barth) eventually published counterexamples. But even before that there had been difficulties in understanding what certain "theorems" were even supposed to *mean*, since expressions like "general position" were used that had no accepted rigorous definition. This amounted to another, very conspicuous case of sloppy generic reasoning, relying on results known to hold only "generally" and not "universally." Not until the work of André Weil and Oscar Zariski became generally known in the 1950s, did the field begin to be put on a fully rigorous basis.[25]

---

[24] The matter is too technical to be gone into here, but a nice popular account of the problem of rigor in connection with maximization and minimization problems, and in particular, in the proof that of all figures with a given perimeter or circumference the circle has the largest area, can be found in chs 21 and 22 of Rademacher and Toeplitz 1966.

[25] Full rigorization came, if anything, even later to the subbranch known as "enumerative geometry," though the problem of setting work in this area on a rigorous basis had already been raised by David Hilbert in the fifteenth of his list of 23 problems at the beginning of the twentieth century. See Kleiman 1974.

Around the turn of the century and even later, not only did non-rigorous methods continue to be used in various branches of mathematics, and even some subbranches of analysis beyond the calculus, but whole *new* branches of mathematics, including even some new subbranches of analysis, were still being introduced on a non-rigorous basis. Two examples are provided by the work of Henri Poincaré, who was no marginal figure, but the leading mathematician of the period.

One of these, *dynamical systems theory*, is well known to the general public, though perhaps not under that name, though unfortunately mainly through hype surrounding two of its subbranches, the apocalyptically named *catastrophe theory* and *chaos theory*. A crucial result in the theory is the *Poincaré recurrence theorem*, which has important applications in mathematical physics, and is the starting point for a whole subbranch, *ergodic theory*. The theorem, applied to standard physical models of the earth–moon–sun system in celestial mechanics, or the atoms in a box of gas in statistical mechanics, tells us, modulo the physical assumption that the model is a good one, that under certain very general conditions, such a system is bound to return again and again, after long intervals of time, if not exactly to its initial position, then at least to one very close to it.

A fairly elementary account is given in Oxtoby (1980, 65–9), where the author remarks that Poincaré's original proof involved considerations of "probability" of which no one at the time could give any rigorous account, but that when read against an adequate background today, which is to say, post-Kolmogorov, "his argument is perfectly sound." This situation illustrates an important general feature of rigorization. A famous short story of Jorge Luis Borges tells of one Pierre Menard, who undertook the impossible task of writing *Don Quixote* in the twentieth century. Borges quotes a passage on history from Menard's work that is word-for-word the same as a passage from Cervantes, and notes how much deeper is the meaning of such words when written by an author coming after, rather than before, William James. So it is with mathematics: Often a text almost word-for-word the same as some older piece of non-rigorous mathematics will have a different meaning, and a rigorous one, merely because it comes after, rather than before, a rigorous account of its background presuppositions.

There is a certain irony, however, in this case of Poincaré's probabilistic reasoning, in that Kolmogorov's rigorization of probability emerged from measure theory, which in turn originated from Henri Lebesgue's efforts to extend integral calculus to cover pathological functions of the kind that figure in Weierstrass's counterexamples, while Poincaré himself tended to disparage concern with such counterexamples. A much-quoted remark of his runs in free translation as follows:

It used to be that when a new function was invented, it was with some practical goal in view; nowadays, they are invented for the sole purpose of finding fault with our forefathers' reasoning, and nothing will be got out of them but that.[26]

Doubtless in some part through Poincaré's influence, such attitudes were common among French analysts, and made trouble for Lebesgue, who wrote that

as soon as I would try to take part in a mathematical discussion, there would be an analyst to tell me, "This can't interest you, it's about functions having derivatives," and a geometer to repeat in his language, "We're concerned with surfaces having tangent planes."[27]

Another branch of mathematics founded by Poincaré is *algebraic topology* and here also his pioneering work—from which emerged the *Poincaré conjecture*, only recently proved—though enormously valuable, was something less than a paradigm of complete rigor. In particular, Poincaré did not always distinguish as clearly as one now would the perspectives of *combinatorial topology* and *differential topology*.[28]

Error in Non-Rigorous Mathematics

All this brings us back to the question posed earlier but postponed, the question of *why* mathematicians ever undertook so ambitious a project as full rigorization. But if the lack of documents makes it difficult to say

[26] *Autrefois, quand on inventait une fonction nouvelle, c'était en vue de quelque but pratique; aujourd'hui, on les invente tout exprès pour mettre en défaut les raisonnements de nos pères, et on n'en tirera jamais que cela.* Poincaré 1920, 132–3.
[27] ... *dès que j'essayais de prendre part à une conversation mathématique il se trouvait un Analyste pour me dire: «Cela ne peut vous intéresser, il s'agit de fonctions ayant une dérivée», et un Géomètre pour répéter en son langage: «Nous nous occupons de surfaces ayant un pan tangent.»* Lebesgue 1922, 13–14.
[28] The full story, involving what was known as the *Hauptvermutung* or *chief conjecture*, and its refutation by John Milnor's example of the "exotic seven-sphere," and subsequent developments down to the time of this writing (when a purported proof that "not every manifold is triangulable" has been posted but not yet vetted by experts) is far too technical to be gone into here, though it is a fascinating one for the reader with sufficient mathematical background.

why the ideal of rigor was first adopted in antiquity, the superabundance of documents makes it almost equally difficult to say why in the nineteenth century it was finally decided to undertake an unprecedentedly serious attempt to realize that ideal. There were so many mathematicians involved, working in so many different branches of mathematics, at so many different centers of research, during so many different decades, and with so many different distinguishable motives, that it is extremely difficult to cite any single factor as definitely *the* reason the project of rigorization was launched—and correspondingly easy to point to multiple factors as possibly *among* the motivations for the project.

The most obvious motivation for rigorization was perhaps the perceived greater reliability of rigorous methods. Whatever methods of discovery stand behind the results reported without explanation in ancient Near Eastern materials, for instance, were clearly not fully reliable, since some of the mensuration formulas arrived at were wrong. Or at least, they were only approximate, and were given without any clear indication that they were recognized as being so. In particular, in all really early cases where the correct formulas involve the constant π, some rational approximation was used, not always as poor as the biblical value $\pi = 3$ (implicit in 1 Kings 7: 23–6), and perhaps good enough for low-tech applications, but nonetheless still not exact. But the unreliability of non-rigorous methods was apparent not only in these cases and others where no supporting arguments were offered—which has happened not only in ancient but in modern times, since some (though not many) of Ramanujan's formulas were incorrect—but also where non-rigorous supporting arguments have been offered.

Analogical extrapolation, and the related method of inductive generalization, for instance, are known to be unreliable. In particular, though a generalization about natural numbers may be verified by calculation up to some very large number indeed, such computations never amount to a rigorous proof, and sometimes suggest outright false results. Let me mention just two of the many cases that tend to get cited in this connection, ones for which the size of the smallest counterexample, the first time the generalization has an exception, and the conjecture goes wrong, is especially large.

One famous case is the *Mertens conjecture*, named for Franz Mertens, who put it forward in the 1890s. It concerns the Möbius function μ, a basic notion introduced in the early pages of any textbook of number theory. The *fundamental theorem of arithmetic* tells us that every whole number greater than 1 can be written as a product of primes, uniquely except for the order of multiplication. The value of the function $\mu(n)$ is taken to be 1 by convention for $n = 1$. It is taken to be 0 if n has any *repeated* prime factor, as in the case of $12 = 2 \cdot 2 \cdot 3$ or $18 = 2 \cdot 3 \cdot 3$, and otherwise is +1 if the number of prime factors is even (as with $6 = 2 \cdot 3$ or $10 = 2 \cdot 5$ or $15 = 3 \cdot 5$) and −1 if the number of prime factors is odd (as with primes like 2 and 3 and 5 themselves, or $30 = 2 \cdot 3 \cdot 5$). The *Mertens function* $M(n)$ is the sum of the values of $\mu(m)$ for m from 1 to n, and the Mertens conjecture is that the absolute value of this function is always less than $\sqrt{n}$. The reader who tries computing a few examples will not find the conjecture to fail. It does, however, eventually fail, though it is known that the smallest counterexample (which has not been explicitly computed) must be greater than 10^{14}.

Another famous case, a bit more technical in its statement, is the generalization that $\pi(x)$, the number of primes less than a given number x, is always less than $\text{li}(x)$, the so-called logarithmic integral function of x. Here, J. E. Littlewood showed that such a conjecture is false, despite its computational verification for what was for the pre-computer era an impressively large range of values of x.[29]

I have alluded already to a number of examples involving other forms of non-rigorous supporting argument: Thus generic reasoning can lead to a bogus "proof" that $0 = \frac{1}{2}$, and spatiotemporal intuition to a fake "demonstration" that a continuous function must be differentiable at all but finitely many points in any given interval. Perhaps the most serious objections to spatiotemporal intuition arose, however, not in connection with cases where it leads to results that are wrong, but in connection with cases where it leads to results *on whose rightness or wrongness mathematicians eventually decided that it was not for them to pronounce.*

Non-Euclidean Geometry and the Division of Labor

Such is the situation in connection with non-Euclidean geometry, whose development is a factor invariably mentioned by commentators in connection with motivations for rigorization in general and the

---

[29] The first upper bound on the size of the smallest counterexample, obtained by Stanley Skewes and accordingly called "Skewes' Number," was vastly larger than almost any natural number specifically considered in earlier mathematics. Even today the smallest counterexample remains unknown, though the upper bound has been substantially reduced, to something under e^{400}.

"degeometrization" or "arithmetization" of analysis in particular (though there is something of an historical mystery here, since the degeometrization process was well under way before non-Euclidean geometry became widely known of in the mathematical community, the pioneering work in the field having been mainly left unpublished or published in obscure venues).

The development of non-Euclidean geometry (and the related, nearly equally important development of four- and higher-dimensional geometry) is itself a long story. Fortunately, there will be no need to recall anything more than a few highlights here. The story began with there being grafted onto the ideal of rigor, by some time in late antiquity, a further requirement concerning the unproved results or postulates and the undefined notions or primitives, namely, that the former should be obviously or "self-evidently" true, and the latter obviously or "self-evidently" meaningful.

The fact, discussed earlier in connection with Propositions 1 and 4 of book 1 of the *Elements*, that Euclid leaves out from his official list of postulates any number of assumptions that would be needed to make his work fully rigorous, left his fifth acknowledged assumption, the *parallel postulate*, in an awkwardly exposed position, as the *only* acknowledged assumption that seemed to lack "self-evidence." Euclid himself implicitly recognized there was something special about this particular postulate, since he does as much as he can without it in the early parts of book 1, before beginning to use it (as he must) to get the results about the sum of the angles in a triangle, and so on. This is one feature of his work still admired even by those who fully recognize its shortcomings.

Any number of later writers, from ancient to modern times, attempted to prove the parallel postulate, without accomplishing more than replacing it with some other postulate, usually one not found significantly more evident by many beyond the proposer himself. The best known of these alternative, equivalent assumptions is perhaps *Playfair's postulate*: Given a line and a point not the line, there is exactly one line parallel to the given line through the given point. Some, notably Girolamo Saccheri in his *Euclid Freed from Every Flaw*, attempted to prove the parallel postulate or a known equivalent by what is called *reductio ad absurdum*, which is to say, by assuming the postulate fails and deducing a contradiction under that assumption. In this way the consequences of the negation of the postulate came to be extensively explored.

Now there are two ways in which the parallel postulate might be assumed to fail, which in Saccheri take the form of what he calls the hypothesis of the obtuse angle and the hypothesis of the acute angle, these being equivalent respectively to the alternatives to Playfair's postulate assuming (1) that there is *no* parallel through the given point; or (2) that there are *many* parallels through the given point. From assumption (1) one gets *elliptical* geometry, in which the sum of the angles in a triangle is always more than two right angles, and the bigger the triangle the bigger the excess. Under assumption (2) one gets *hyperbolic* geometry, in which the sum of the angles in a triangle is always less than two right angles, and the bigger the triangle the bigger the defect.

Saccheri pursued the consequences of his obtuse and acute hypotheses in the hopes of arriving at a contradiction. He found one in the case of the obtuse angle hypothesis, though it depends on an interpretation of Euclid's postulate about the possibility of extending any line segment, according to which the postulate presupposes or insinuates the infinity of space. (For the obtuse angle or no-parallels assumption of elliptical geometry ends up implying that space is finite, and a line extended long enough doubles back on itself.) In the case of the acute angle or many-parallels assumption of hyperbolic geometry, Saccheri did not accomplish even so much, but merely reached a conclusion that, though not amounting to a genuine reductio, he considered repugnant.

Both elliptical and hyperbolic geometries are in fact consistent, and the consistency of both forms of non-Euclidean geometry was eventually rigorously established (assuming the consistency of Euclidean geometry itself) by producing models of the two geometries within Euclidean geometry. Elliptical geometry is modeled by *spherical geometry*, the "plane" being the surface of a sphere, and "lines" being great circles on it. Hyperbolic geometry is modeled by the so-called *Poincaré disc*, the "plane" being the region inside a circle, and the "lines" being those arcs of circles that meet the boundary circle at right angles.[30] And even before

---

[30] Strictly speaking, the spherical geometry model is a model of what is called *double* elliptical geometry, since in it any two lines have *two* points of intersection; a different model—the so-called *projective plane*, which is just a bit harder to describe—is used for *single* elliptical geometry. The Poincaré disc model is widely familiar, even to non-mathematicians from M. C. Escher's *Circle Limit* graphics. By the way, Smith 1929 gives selections from Saccheri (351–9), as well as from Bolyai (371–88) and Lobachevsky (360–9), both to be mentioned shortly.

proofs of consistency by modeling were given, a conviction of the consistency of non-Euclidean geometry had been arrived at by some of those who wandered into this strange area of thought. Two of them, Janos Bolyai and Nikolai Lobachevsky, were bold enough to publish their work.

Another of them, Gauss, who had more of a reputation at stake, withheld his work from publication, it is said for fear of conflict with philosophical followers of Kant. Now actually the consistency of non-Euclidean geometry merely confirms one half of Kant's view of Euclidean geometry, which was that it is what he called "synthetic" rather than "analytic," which is to say that it does not consist merely of logical consequences of definitions. What came to be questioned was the other half of Kant's view about Euclidean geometry, namely, his claim that it is what he called "a priori" rather than "a posteriori," which is to say, that its principles are not inferred inductively from sense-experience but are somehow knowable by pure thought. The consistency of non-Euclidean geometry does not logically imply that Euclidean geometry is a posteriori, but almost as soon as the consistency of non-Euclidean geometry was taken to heart by mathematicians, some of them began questioning whether the Kantians had any really convincing grounds for their claim that Euclidean geometry is a priori.

Gauss expressed his rejection of the claim that Euclidean geometry is a priori by saying that it has the same character as mechanics. To say this is not to say that it is false, but it is to say that its truth or falsehood remains open to empirical test, and a question ultimately for physicists and not pure mathematicians to decide. For instance, one might survey three mountaintops and measure whether the angle-sum is 180°. Gauss is said to have done this with the peaks at Hoher Hagen and Grosser Inselsberg and Brocken, during the course of a survey of Hanover today commemorated by the "Gaussturm" observation tower at Dransfeld, near Göttingen, where Gauss taught.

Actually, the situation is a little more complicated. This is in part because—as Gauss already realized—the non-Euclidean geometries imply that space is *locally* and *approximately* Euclidean, so that its non-Euclidean character might fail to be recognized experimentally because our observations are not extensive or exact enough. More seriously, no geometrical theory implies any empirical predictions all by itself and without any physical auxiliary hypotheses. Even the proposed surveying experiment assumes that light travels in straight lines, and so involves not only geometry but also optics, which is to say, electromagnetic theory. It has turned out that geometry is likewise inseparable from gravitational theory.

The hypotheses about the geometry of space—or rather, of space-time—that emerged with special relativity's treatment of electromagnetism and general relativity's treatment of gravitation are further from traditional Euclidean geometry than were its early nineteenth-century elliptical and hyperbolic rivals. And neither general relativity nor quantum field theory, which provide the best available theories of gravity and of electromagnetism and kindred forces, respectively, is wholly acceptable: Relativity theory needs a quantum correction and quantum theory needs a general-relativistic correction. What the geometry associated with the ultimate "theory of everything" will be—supposing one is ever arrived at—remains in doubt. And so the story is not yet over, and there is as yet not agreement even as to the number of dimensions required, with 11 and 26 among the candidates most discussed.

The part of all this story that is most pertinent to present purposes is a distinction, implicit already in Gauss's standpoint, that came increasingly to be made, between *mathematical geometry* and *physical geometry*. The role of mathematics became that of exploring various mathematical "spaces," Euclidean and non-Euclidean, two- and three- and four- and higher dimensional, while the role of physics was to decide which of these was most appropriate as a model of the physical space in which we live and move, and whose remoter reaches astronauts explore. In particular, whether anything like the parallel postulate holds of the *physical* space in which we live and move is not a matter for *mathematicians* to decide.

Of equal importance with the realization that it must be left to physics to judge which mathematical space is appropriate to model physical space was the realization that a mathematical space not appropriate as a model of physical space *might very well serve as a model of a different kind for some other empirical phenomenon*. An example is provided by the use of $6n$-dimensional space as the so-called *phase space* of an n-particle system in physics, which is the setting for the Poincaré recurrence theorem cited earlier. Another example is the representation of economic data by points in a multidimensional Euclidean space. General results about, say, Euclidean spaces of arbitrary dimension—even of dimensions much higher than 11 or 26 or any number that has been hypothesized for physical space—may turn out to have quite unanticipated applications to natural, or even social, sciences. Such is the case, with the result that two convex regions in an $(n + 1)$-dimensional space are separated by an n-dimensional subspace, so that two convex figures in the plane can be separated by a line, two convex bodies in space by a plane, and analogously in higher dimensions.

In the new way of working that emerged, there is a new division of labor. The mathematician studies some mathematical space or system. The empirical scientist proposes hypotheses to the effect that it models in some way some natural or social phenomenon, and then applies mathematical results about the mathematical space or system, in conjunction with the basic hypothesis that it models the natural or social phenomena in question, to make empirical predictions. If the predictions fail, the empirical scientist's hypothesis must be modified or abandoned, and some other model sought, *but the mathematical results remain*, perhaps to be recycled and reused some day in modeling something else.

This new way of working, however, clearly presupposes that the mathematician, in obtaining theorems, does not at any point make use of the empirical scientist's hypothesis that the mathematical space or system at issue well models some familiar natural or social phenomenon. Nothing may be assumed in proofs about the mathematical space or number system simply because the corresponding statement seems to be correct about the empirical phenomenon it is supposed to model,[31] if we are to be sure that the mathematical result still holds even if that supposition turns out to fail. In particular, spatiotemporal intuition must not be involved in proofs. Only rigorous proofs deducing every theorem that is claimed from explicit assumptions definitive of the mathematical space or system in question can be admitted, if mathematics and empirical science are to be able to work together under the kind of division of labor just described: the one alluded to earlier in contrasting the roles of "mathematical physicist" and "theoretical physicist."

This point about the role of rigor in the division of labor between mathematics and the other sciences has recently been isolated and emphasized by Frank Quinn (2012), whom we will encounter again in Chapter 2 as a participant in a recent explicit debate about rigor among professional

---

[31] This is so despite the frequent utility of such assumptions in heuristically suggesting conjectures to be proved, as in the famous case of the use of soap films and soap bubbles in connection with what is known as *Plateau's problem*, as in Newman 1956c (ii. 891–909).

mathematicians, a debate of a kind that is really quite rare. Quinn offers a metaphor according to which, under the new dispensation, with its division of labor, rigorous mathematics is like the bony endoskeleton to which the musculature of empirical science is attached, whereas under the old dispensation, where mathematics was thought to be directly about certain aspects of the world around us, it was more like a chitinous exoskeleton. Creatures with endoskeletons can grow to much greater sizes than creatures with only exoskeletons, and Quinn warns that without continued due attention to rigor in mathematics, whose value funding agencies tend to underestimate, science will have to "go back to being a bug." Like another Gregor Samsa, science may find itself metamorphosed from vertebrate to arthropod.

Error in Rigorous Mathematics

Thus in one way or another the appeal to spatiotemporal intuition, as much as the appeal to inductive generalization, fell into a certain degree of disrepute. Nonetheless, as has already been mentioned in connection with the Euler series, for instance, non-rigorous methods can sometimes be quite convincing, while inversely, rigorous methods are by no means infallibly reliable. There is no certainty in human life, and there can be errors in mathematics that aspires and claims to be rigorous as much as in mathematics that frankly is not.

The cautious attitude towards brand new results that Hume described is justified by cases where apparent "proofs" have fallen through. Much of this erroneous would-be rigorous mathematics—including all the papers that get rejected by editors because a referee finds a flaw in the proof of the main theorem—won't be visible to one browsing the university library. Nonetheless, one does not have to flip through too many back issues of too many journals before one turns up a retraction. It is not without reason that the Clay Foundation will pay out one of its million-dollar "Millennium Prizes" only a couple of years after the purported solution has appeared in a peer-reviewed journal, to allow public scrutiny of the purported proof.

There have been some notorious examples of purported rigorous proofs that have turned out to be fallacious, in some cases after being accepted for a decade or more. To mention one example not involving any living mathematicians, Alfred Kempe's 1875 "proof" of the so-called four-color theorem (saying roughly that no more than four colors are required to color a map in such a way that two adjoining countries are always of different colors) was exposed as deficient by Percy John Heawood only in 1890. Proofs depending on calculations, in particular, are especially likely to go wrong if the calculations are long and complicated enough. And so the mere desire to avoid error cannot be the whole story of the motivation for rigorization.

In approaching such issues as the question why mathematicians eventually went in for rigor in a serious, sustained way, one must always have in the back of one's mind the following remark from Weyl:

> the prevailing attitude has been one of resignation. The ultimate foundations and the ultimate meaning of mathematics remain an open problem; we do not know in what direction it will find its solution, nor even whether a final objective answer can be expected at all. "Mathematizing" may well be a creative activity of man, like music, of primary originality, the products of which not only in form but in substance are conditioned by the decisions of history and therefore defy complete objective rationalization. (Weyl 1949, 219)

This we may call Weyl's Null Hypothesis.[32]

---

[32] Weyl, before this passage, has just been writing apropos of the debates over the direction of mathematics between Hilbert and Brouwer, to which we will come in Ch. 2. In these debates, Weyl had sympathized with the losing, Brouwerian side, which may partly explain the somewhat fatalistic tone of the remark.

2

Rigor and Foundations

The Axiomatic Method and Ideal Elements

From a larger point of view, the introduction of non-Euclidean spaces, and then various multidimensional spaces and more, is merely the manifestation in geometry of a generalizing tendency happening across all nineteenth-century mathematics. The same tendency was evident outside geometry, as in algebra with, for instance, Hamilton's introduction of his *quaternions*, and then Arthur Cayley's introduction of his *octonions*. In the algebraic case, it is quite clear that habits of thought acquired while working with more familiar number systems do not all carry over to the new systems, an observation that points towards the need for greater rigor. Most famously, the rule for multiplying one quaternion $a + bi + cj + dk$ by another does not obey the commutative law, since $ij = k$ but $ji = -k$; and things are worse with the octonions, where even the associative law breaks down.

It should be remarked, however, that so long as one is dealing with unfamiliar spaces and number systems and whatever one at a time, one does eventually acquire, after working for some time with any one of them, habits of thought—call them "intuitions" in a non-Kantian, non-spatiotemporal sense, if you will—appropriate to that one space or number system. Rigor may then well lapse after a while, in the sense that one comes to rely on such hunches bred of familiarity acquired through experience with how the once-novel space or number system works, unless the surrounding mathematical culture is one in which observance of the requirements of rigor has become ingrained.

A stronger impetus towards rigor arises where one wants to deal simultaneously with a whole batch of unfamiliar mathematical spaces or number systems or whatever, taking in all those satisfying some specified conditions, even though there may be unsurveyably many that satisfy them.

In such a case, it seems, nothing less than rigorous deduction from the stated conditions as postulates or unproved starting points will do. And in the nineteenth century talk of the "axiomatic method" less often meant the method imperfectly approximated by Euclid, that of rigorous deduction from postulates about a single space or number system, than the method of treating whole batches of examples at once, some general *kind* of space or number system, by restricting oneself to rigorous deduction from the conditions satisfaction of which *defines* the kind of space or number system in question. (For this reason talk of "the axiomatic method" was often accompanied by obscure talk about "implicit definitions.")

Thus, *generalization* motivated rigor in both geometry and algebra. A kind of generalization was at work in analysis as well. Already in the eighteenth century some mathematicians adopted something like the modern definition of *function* from and to the real numbers as any association whatsoever, however arbitrary it may be, of a real number as output or value to any given real number as input or argument. Those who assented verbally to such a definition did not always stick to it in practice, however, insofar as they might sometimes state theorems for "all" functions, while tacitly assuming in the proof that every function of interest is continuous or differentiable, or can be represented by some analytical expression, or something of the sort. Unclarity about such points was part of what was involved in a famous three-cornered debate among Jean le Rond d'Alembert, Euler, and Daniel Bernoulli over the mathematical modeling of a vibrating string.[1]

But gradually, in the nineteenth century, it was realized that, if one insists on full generality, and adheres to the very broad concept of function, then one must admit many badly behaved or pathological functions, and one must state one's theorems carefully, making any assumptions of continuity or differentiability needed explicit: that the function is continuous, that it is differentiable, that it is differentiable and with a derivative that is continuous, that is twice differentiable or differentiable with a derivative that is in turn differentiable, that it is smooth or has derivatives of all orders, or that it is analytic or representable by a series. The Weierstrass function (everywhere continuous, nowhere differentiable) is but one example of a pathological function satisfying one of these assumptions but not another.

---

[1] The account in Kline 1972, ch. 22, sect. 2, of this important incident is somewhat technical, but can be followed with a knowledge of little more than freshman calculus.

Another and worse case is presented by Lejeune Dirichlet's example of the function taking the value one for rational arguments and zero for irrational arguments. Rigorization was associated with a widening of the class of functions given serious recognition, a somewhat different kind of generalizing tendency from that manifest in geometry and algebra.

To cite a desire to generalize as a motive for insisting on rigor only raises another question, however, the question why generalization itself was insisted upon, the question of what motivated *it* in turn. And here, at least in geometry and algebra, a good part of the reason has to do with a certain indirectness of method characteristic of much of modern mathematics, in which an original space or number system of primary interest is investigated by considering a related auxiliary space or number system, which becomes an item of secondary interest on account of what information about it may imply about the item of original interest; further auxiliaries to auxiliaries may then be introduced in turn, and become objects of tertiary interest.

In several early and important cases, the auxiliary simply consists of the original space or number system expanded to a larger space or number system by adding 'ideal elements.' A principle much emphasized by Hilbert, but familiar already well before his time, is that adding ideal elements is often the shortest and most efficient way, and may even for some time remain the only known way, to reach results about the space or number system of primary interest.

A very early example is provided by the "imaginary numbers" that, added to the real numbers system, give us the complex number system. To cite the earliest great achievement of modern mathematics, in the so-called *casus irreducibilis* of a cubic equation (where there are three distinct real, non-rational solutions), to give a so-called solution by radicals (a formula involving only addition, multiplication, subtraction, division, and extraction of roots that expresses the solutions in terms of the coefficients), one must at some point take a detour through the complex numbers.

Another example is provided by the theorem that any prime number that leaves a remainder of one on division by four can be written as a sum of two squares, as with

$$5 = 4 \cdot 1 + 1 = 2^2 + 1^2$$

$$13 = 4 \cdot 3 + 1 = 2^2 + 3^2$$

Though various proofs are known, one that is used in many textbooks today takes a detour through consideration of the "Gaussian integers" or complex numbers of the form $a + bi$ with a and b both integers.

After the nineteenth-century development of *complex analysis*, or calculus for functions whose arguments and values include imaginary numbers—the single most important manifestation of the generalizing tendency in analysis—the number of results about the real domain involving a detour through the complex domain became enormous. To cite one celebrated example, *Chebyshev's theorem*, a.k.a. *Bertrand's postulate*, which tells us that there is a prime number between any natural number greater than one and its double, was originally proved using integration of complex-valued functions.

Though I have been emphasizing the introduction of ideas from analysis into number theory as an indirect means of attacking traditional number-theoretic problems, themselves *statable* without use of analytic terminology, the introduction of ideas from analysis into number theory also vastly enriches mathematics by opening up a large domain of questions that cannot be formulated in purely number-theoretic terms, and what is called *analytic number theory* became a growth area in nineteenth-century mathematics. New issues include especially questions about the distribution of primes, of which we have seen one example already (the false conjecture, suggested inductively or experimentally by computational verification of many cases, that $\pi(x) < \mathrm{li}(x)$ for all x). And there are many others: the so-called *prime number theorem* (according to which $\pi(x)/x$ approaches $1/\ln(x)$, where ln is the natural logarithm function, as x approaches infinity, a result it took the better part of a century to prove), and the *Riemann hypothesis* (alluded to earlier, and a little too complicated to state here), which remains an open question.

But to repeat, the introduction of analytical or algebraic auxiliaries also leads to *proofs* of many theorems whose *statements* involve only natural numbers. The proof by Andrew Wiles (partly in joint work with Richard Taylor) of Fermat's conjecture is a spectacular recent example. Sometimes "elementary" proofs are subsequently discovered, where "elementary" means working strictly with the natural number system, *bringing in no auxiliary apparatus*. Such is the case, for instance, with Chebyshev's theorem, where an "elementary" proof was found by Paul Erdös. Such "elementary" proofs tend to be more rather than less difficult compared with the first proofs, involving auxiliaries, to be discovered. Whether there is an "elementary" proof of Wiles's theorem is at present unknown.

Another fairly early example of the introduction of ideal elements is provided by the "points at infinity" and "line at infinity" that added to the Euclidean plane or two-space give us the projective plane or two-space. Each line is supposed to contain a "point at infinity," with parallel lines sharing the same one; the points at infinity taken together make up the "line at infinity."[2]

The first thing that the introduction of such auxiliaries does for us is to permit the more compact formulation of various results. This advantage is illustrated by the case of *Desargues's theorem*. Two triangles ABC and A'B'C' are said to be in *perspective* if the three lines through corresponding vertices (which is to say, through A and A' and through B and B' and through C and C') meet in a single point P. The theorem states that on the hypothesis that the triangles are thus in perspective, if each of the three pairs of lines extending corresponding sides (which is to say, the two lines through A and B and through A' and B', and the two lines through B and C and through B' and C', and the two lines through C and A and through C' and A') meets in a point, then the three intersection points K and L and M all lie on a single line.

In Euclidean geometry, the foregoing is just the first case of the theorem. A second case says that the same conclusion holds on the alternate hypothesis that the three lines through corresponding vertices are all parallel. A third and a fourth case say that on either hypothesis, if two of the three pairs of lines extending corresponding sides (say the two through A and B and through A' and B' and the two through B and C and through B' and C') meet, while the third pair of lines extending corresponding sides (the two through C and A and through C' and A') are parallel, then the line through the two points of intersection K and L is also parallel to the latter pair. A fifth and a sixth case say that, on either hypothesis, if two of the three pairs of lines extending corresponding sides are parallel, then so is the third. (It will be worth the effort for the reader not previously acquainted with these matters to think through the six cases, drawing diagrams.)

---

[2] Extracts from the account in Kline 1972, ch. 14, appear in Newman 1956c, i. 622–41, including diagrams illustrating Desargues's theorem and others to be mentioned shortly. The same part of that anthology contains several selections pertaining to topology, to be discussed later in this chapter.

In the projective formulation, by contrast, there is only one case, since two lines *always* meet in a point. The two hypotheses considered on the Euclidean formulation amount, when considered from a projective point of view, to the hypotheses that the three lines through corresponding vertices meet in an ordinary point P and that they meet in a point at infinity P. The three conclusions are that the three points of intersection of K and L and M are all ordinary points, lying on a single line, or that two of them (say K and L) are ordinary points and the third (namely M) is the point at infinity on the line through them, or that all three are points at infinity, lying on the line at infinity. Thus, a single projective formulation encompasses a half-dozen Euclidean formulations.

The second thing the introduction of the auxiliaries does is to introduce a symmetry into the basic postulates, since just as every two distinct points determine a unique line, so also every two distinct lines determine a unique point. A result is the *principle of duality*, according to which any theorem rigorously deduced from the axioms remains true if the words "point" and "line" are interchanged. This interchange makes no difference in the case of Desargues's theorem. Popularizations illustrate the substantive use of duality by the case of *Pascal's theorem* and *Brianchon's theorem* about hexagons, but the details of their statements need not detain us here.

Consideration of the complex number system and of projective spaces (not only in dimension two, but in higher dimensions also) are just two early examples of the method of introducing ideal elements. The nineteenth century introduced many more, a couple of which deserve special mention. The very terminology "ideal element" derives from Ernst Kummer's "ideal divisors," auxiliaries introduced mid-century in his study of special cases of the Fermat conjecture. Number theory towards the *fin de siècle* saw another kind of ideal element in Kurt Hensel's *p-adic* numbers. Again, technical details need not concern us.

The Unity of Mathematics

The auxiliaries introduced are not always ideal elements added to the given space or number system to produce an extended one, as in the examples considered so far, but are sometimes of a quite different character. Such is the case with the apparatus used in Wiles's proof, for instance. A fine early example is provided by *Galois theory* (the usual goal of an undergraduate course in abstract algebra).

Here, the primary interest is in finding, in the complex numbers, the *roots* of a polynomial $f(x)$, which is to say the *solutions* to the equation $f(x)$ = 0, especially in the case where the coefficients of the polynomial come from the rational numbers. Towards this end, one associates with the polynomial an auxiliary system consisting of *permutations* of those roots. These form something like a new number system when one construes *composition* of two permutations (first performing one, then performing the other) as a kind of "multiplication." The resulting theory unifies a great variety of results pertaining to classical problems, illustrating the fact that auxiliaries really are of interest not just for their own sake (though they are of interest for that), but for their bearing on traditional questions. This fact is sufficiently important that it is worth underscoring it by elaborating a bit about the applications of Galois theory that illustrate it, which fortunately can be done without needing to describe the details of Galois theory itself.

Euclid, in his first proposition, in effect shows how to construct with straightedge and compass a regular (that is, equilateral) triangle,[3] and much later in his work shows also how to construct a regular pentagon. Gauss showed how to construct a regular 17-gon. Now if one wants to show some figure is constructible, it suffices to indicate the construction (and prove that it works). To show that some construction is *impossible*, to show for instance, that a regular heptagon can *not* be constructed with straightedge and compass, one needs some kind of analysis of what constructibility amounts to. The introduction of coordinate methods made it possible to formulate such an analysis, to give an algebraic criterion for constructibility, turning geometric construction problems into algebraic problems in the theory of equations. And Galois theory becomes relevant by being applicable to such problems.

One application is precisely to show that the regular heptagon (or 11-gon or 13-gon, and so on) is not constructible. Other applications show that the ancient construction problems of the *duplication of the cube* (construction of a cube of exactly twice the volume of a given cube) and the *trisection of the angle* (division of the angle into three equal angles)

---

[3] I say "in effect" because Euclid does not mention any particular instruments, and the constructions he allows (connecting two points by a line or extending a given line, and describing a circle) can be carried out without straightedge or compass by the Egyptian surveyors' method of using a cord pulled taut.

are impossible—something that had already been concluded by various academies in the eighteenth century, which refused to accept papers from unknowns purporting to give solutions to such problems. Another application takes us a long way towards, though not all the way to, a similar conclusion about *squaring the circle* (finding a square equal in area to a given circle). Another application shows that the Renaissance Italian feat of providing solutions in radicals to cubic and quartic equations cannot be pushed further: There is no general solution in radicals to quintic equations.

A different example of the use of auxiliaries is provided by geometry, where many problems concerning the Euclidean plane or higher-dimensional Euclidean spaces are addressed by considering various *transformations*, or mappings of points to points. For instance, in the plane the *rigid motions*, the transformations that preserve distance, can be shown to consist solely of translations, rotations about a point, and reflections in a line, and combinations thereof. Information about transformations can be used to establish facts about the figures transformed, facts whose statement does not explicitly mention the transformations considered as auxiliaries. Euclid's use of superposition, already discussed, amounts to an early, implicit example, but much more is possible when the transformations are made the object of explicit study.

Focusing on transformations also opens up a new realm of questions, particularly concerning *symmetry*. For instance, it can be shown that among repeating wallpaper patterns only fourteen types of symmetry are possible. An analogous result in three dimensions, with 230 possibilities, has important applications in crystallography, a minor illustration of the fact that the importance of the new method of studying transformations extends from pure to applied mathematics. A major illustration of the same fact is provided by the contrast between *Galilean* and *Lorenz* transformations, which are symmetries not of physical objects but of physical laws, distinguishing classical from special-relativistic physics.

Felix Klein famously proposed, in what is known as the *Erlanger program* (after the town, Erlangen, where it was to be announced), to bring order to the chaos of contrasting mathematical geometries by using the transformation idea. For instance, ordinary Euclidean geometry studies properties preserved under rigid motions and *dilatations* (uniform expansions, multiplying all distances by a constant factor). By contrast, other branches of geometry study properties preserved under wider classes of transformations. An important case is that of topology, formerly called *analysis situs*. So far as it applies to the plane, it includes such results as the four-color theorem (mentioned earlier), Euler's analysis of the famous Königsberg bridges problem (it is not possible to cross all seven bridges in a single route without crossing at least one of them more than once), the Jordan curve theorem (any closed path which does not cross itself divides the plane into an inside and an outside), and many more. It may be characterized, on Klein's scheme, as dealing with the widest class of transformations, including all sorts of bendings and stretchings, for which there remain any interesting invariants or preserved properties.

The permutations of roots considered in the theory of equations and the transformations of points considered in geometry and topology were themselves recognized as just two instances of the general notion of a *group*, and this subsumption of several distinct topics into a single theory was one of the first and most important examples of the "axiomatic method" in nineteenth-century mathematics.[4] The development of group theory illustrates not one but two important features of modern mathematics.

First, there is the indirectness of method that I have been discussing. At the start, some traditional or modern problem concerning some traditional space or number system is considered. Auxiliaries are introduced and recognized as constituting an example of some species (such as "group") characterizable by axioms. General results from the theory of such items (such as "group theory") are applied to establish results about the auxiliaries. And finally, those results are applied to establish results about the original space or number system, solving the traditional problem from which we started.

Second, there is the interconnectedness of the different branches of mathematics, a phenomenon evident since the seventeenth century in the use of coordinate methods, but vastly expanded in the nineteenth

---

[4] Again, Newman 1956c provides a couple of pertinent selections in iii, part IX. The title Newman gives to this part is "The Supreme Art of Abstraction," though the nineteenth-century development of group theory in fact looks very concrete (in a mathematical, non-philosophical sense) compared to such twentieth-century developments as category theory. It is somewhat surprising, giving his fondness for anecdotes, that Newman's introductory commentary omits the tale (which can be found in several variant versions in several different sources) about James Jeans and his supposed advice, fortunately not followed, to the Princeton physics department about its mathematical curriculum: "We may as well cut out group theory. That is a subject which will never be of any use to physics."

century. With the group concept, an idea originating in algebra is applied to geometry. With "functional analysis," ideas originating in geometry or topology are applied to analysis, as functions come to be considered "points" in an auxiliary space, and operations like differentiation and integration come to be considered as "transformations" of that space.[5] And so on across the whole of mathematics. Mathematics is no motley.[6] What appear at the ground floor to be separate pavilions, quite different fields of study (number theory and algebra and analysis and geometry and topology) turn out in their higher stories to have all sorts of connections to each other.

The centerpiece of nineteenth-century mathematics was the development of complex analysis, which has a very different flavor from real analysis (to begin with because the hypothesis of differentiability once implies differentiability any number of times, and representability by a power series). And the centerpiece of complex analysis, and the poster child for the interconnection of different branches of mathematics, was the theory of Riemann surfaces, which, in a way that can be made precise, involving the application of geometric ideas to analysis, turned many-valued functions like the function $f(z) = \sqrt{z}$ (which has two values for any argument except zero, since any such argument has *two* square roots, one the negative of the other) into single-valued functions.

What matters, for purposes of the analysis of rigor, about the two features of modern mathematics I have been stressing—indirectness of method and interconnectedness of branches—is that they have crucial implications for the project of rigorization. For interconnectedness implies that *it will no longer be sufficient to put each individual branch of mathematics separately on a rigorous basis.* By the end of the nineteenth century this latter, lesser task had in fact been more or less accomplished

---

[5] One reason one needs to allow in pathological functions like the Riemann-Weierstrass examples is in order to achieve a certain "completeness," analogous to the completeness of the real number-line, in the "space" of functions.

[6] Talk of the "motley" of mathematics in Wittgenstein's posthumously published *Remarks on the Foundations of Mathematics* (1978) is probably to mathematicians the second most shocking feature of that work, after the insistence that proofs must be "perspicuous"—not an adjective anyone would apply to the 250-page proof of the Feit-Thompson theorem in group theory, for instance. For an overview, see Maddy 1993, who emphasizes Wittgenstein's claim that philosophical clarity about mathematics would inhibit the (to him pathological) growth of the subject. If there was thus much of mathematics that Wittgenstein did not appreciate, mathematicians may take comfort in the fact that he did not appreciate Shakespeare either.

so far as the core subjects of pure Euclidean geometry and of number theory are concerned, in works such as Hilbert's *Foundations of Geometry* and Giuseppe Peano's *Formulaire Mathématique*.[7]

But as we have just seen, by 1900 it had long been the case that *pure* Euclidean geometry is not Euclidean geometry enough, and still less was *pure* number theory number theory enough. Peano treats more material than the mention of "arithmetic" in his title suggests, but makes no direct provision for the use of sophisticated analytical and algebraic methods in number theory, while Hilbert makes no real provision for the use of transformation groups.

To guarantee that rigor is not compromised in the process of transferring material from one branch of mathematics to another, it is essential that the starting points of the branches being connected should at least be compatible. This matters already with the seventeenth-century application of algebra to geometry through coordinate methods. If one is going to use such methods, it really will not do to assume in geometry the so-called *Archimedean axiom*, to the effect that for any two lengths, some whole number multiple of the lesser exceeds the greater—something presupposed in Eudoxus' treatment of proportion—while assuming in algebra the existence of infinitesimals, numbers greater than 0 but smaller than one-half, one-third, one-fourth, and so on.

And it should be emphasized that the application of results from one branch of mathematics to another is by no means just a matter of repeated application of a few familiar, basic connections, such as the use of coordinate methods just mentioned. On the contrary, new connections are constantly being made right down to the present day, though naturally the most recent examples are more advanced and technically sophisticated than familiar, basic ones.[8]

[7] Both works cited do date from the nineteenth century, though the versions listed in the bibliography at the end of this book are from the twentieth. Hilbert 1902 is the English translation by Townsend of the original 1899 *Grundlagen der Geometrie*, written in German. Peano 1901 is the French translation by Peano himself of the original 1895 *Formulario mathematico*, written in Latino sine Flexione, an artificial language of his invention.

[8] They are too advanced and technically sophisticated to describe here, I fear. For the *cognoscenti*, I am thinking especially of the successive introduction of increasingly sophisticated topologies by Zariski, Grothendieck, and others in algebraic contexts. If Zariski is to introduce his topology and then apply results of general topology, it is essential that algebra and topology should rest on assumptions that are at least compatible, unlike an algebra with infinitesimals and a geometry with an Archimedean axiom.

The only obvious way to ensure compatibility of the starting points of different branches is ultimately to derive all branches from a common, unified starting point: The *material* unity of mathematics, constituted by the interaction of its various branches at their higher levels, virtually imposes a requirement of *formal* unity, of development within the framework of a common list of primitives and postulates, if the rigorization project is to be carried to completion.

Meanwhile, the indirectness of method commented upon earlier implies that *the common, unified starting point will have to be such as to make provision for all the types of constructions by which new, auxiliary spaces or number systems or whatever are manufactured out of old, traditional ones.* There are a handful of basic types of constructions that keep being used over and over, in ever more elaborate combinations.

One is the simple formation of *ordered pairs*, which can be iterated to give triples, quadruples, and so forth. Such is the method used to obtain the complex numbers and then quaternions or octonions from the real numbers, for instance. Another method is the formation of the family of all functions or mappings from the points or numbers of a given space or system to other such points or numbers, followed by the separation out from these of special subfamilies, such as the permutations, or the translations or rotations or dilatations that provided group theory with its first examples. Yet another method is the consideration of *equivalence classes.* For instance, calling two integers equivalent if they leave the same remainder upon division by four, the equivalence classes are the *integers modulo four*, which have an algebra (an addition and a multiplication) of their own. This sort of modular arithmetic plays a large role in all post-Gaussian number theory. Expressions like "product" and "power" and "quotient"— not of numbers, but of spaces or number systems, or whatever—came to be applied to such constructions.

To complete the project of rigorization, a framework not only common but commodious would be called for, one accommodating (1) all traditional branches; and (2) all methods for constructing new spaces or number systems or whatever from old, with all their actually existing and all their potential future applications. Actually, under heading (1) it is enough to accommodate traditional number theory, since the methods mentioned in (2) can be used to reconstruct all other traditional spaces and number systems once the natural numbers are available as a starting point: This is what is accomplished by the "arithmetization" or

"degeometrization" project in analysis. This much was more or less realized by late in the nineteenth century, and is reflected in the saying attributed to Leopold Kronecker to the effect that "the Good Lord made the
whole numbers, everything else is human invention."[9] Still, even after this
slight reduction, to accommodate (1) the natural number system; and (2)
formations of products and powers and quotients of systems, in a single
comprehensive system, is a tall order.

Logic and Foundations

So much for the moment about what needed to be done, so far as the scope
of the mathematics that had to be accommodated by a rigorous framework is concerned. The question of what had to be done, in the different
sense of what features a development must have in order to count as genuinely rigorous, needs also to be considered or reconsidered. So far, citing
Aristotle, we have mentioned the need for the derivation by *definition* and
deduction of new notions and results, proximately from older notions and
results, and ultimately from first principles, from primitives, and postulates. We now need to look a little closer at what deduction and definition
themselves amount to, and for the first time take a look at what modern
logic has to say about such matters, and enter into the problem of characterizing the *logical* derivation of new mathematics from older mathematics and first principles.

Logic, as represented in present-day textbooks, does tell us at least
one thing about deduction. A genuine deduction of a conclusion Q from
premises $P_1, P_2, \ldots, P_n$ will show that Q is a *consequence* of the P_i, where
this in turn means that the *logical form* alone of the premises and conclusion guarantees that *if* all the P_i are true, then so is Q. Here "logical form"
refers to the way in which a premise or conclusion is built up using such
logical particles as "not" and "and" and "or" and "if" and "all" and "some."
Similarly, a genuine *definition* of a notion N in terms of $M_1, M_2, \ldots, M_n$ will
make N in effect an abbreviation for some long expression built up from
the M_i and logical particles, and there is perhaps not much more to be said
about definition while operating at the present high level of generality.

---

[9] *Die ganzen Zahlen hat der liebe Gott gemacht, alles andere ist Menschenwerk.* See the
obituary notice in Weber 1893.

But as to deduction, the condition that form alone guarantees the truth of the conclusion given the truth of the premises thus implies that *in any other argument of the same form, if the premises are true, then the conclusion is true.* To put the matter another way, if Q is a genuine consequence of the P_i, then so long as we leave the logical particles alone, we can go through Q and the P_i and systematically replace each item of *nonlogical vocabulary* by a different expression, to obtain some Q' and some P_i', then it will still be the case that if all the P_i' are true, Q' will be true. To put the matter yet another way, if without actually changing any expressions we agree to impose systematically an *unintended interpretation* on each item of nonlogical vocabulary, then it will still be the case that, if all the premises are true, the conclusion will be true.

In the history of mathematical logic, this kind of characterization of logical consequence derives mainly from the work of Alfred Tarski, which is indeed in advance of all previous treatments insofar as it is based on a clear definition of truth, but the general idea that logical consequence is a matter of form is much older.

That he has some such criterion of consequence in mind is what explains why even such a pre-Fregean, essentially Aristotelian logician as Lewis Carroll illustrates points about logical consequence by examples in which the nonlogical vocabulary is far-fetched and fantastical. For instance, one of his most famous problems (Carroll 1897, 124) has as its conclusion "I always avoid a kangaroo." The connection between consequence and form is reflected in the very name "formal logic" for logic in the narrow and strict sense—the subject that Aristotle called "analytics" and his disciples called "organon" and the medievals called "dialectic"—as contrasted with "logic" in the broad and loose sense, in which the label has been used by Kant and Hegel and others. Indeed, one can trace such ideas all the way back to logic's founder, Aristotle, who already uses expressions on the order of "All As are Bs," such as were used earlier in describing the syllogism in Barbara, expressions with letters in place of terms, to indicate logical forms—and this at a time well before letters were used in place of numbers in algebra. The conclusion of a syllogism in Barbara is a consequence of the premises because the form alone guarantees, regardless of the subject matter (regardless of what substantive terms are put in for the letters A and B and C), if the premises are true the conclusion must be true.

Now mathematicians will not be dictated to by logicians. On the contrary, what we know as mathematical logic was in large part developed by Frege and Russell in order to be able to represent, as the logic of Aristotle or Lewis Carroll could not, the forms of argument that mathematicians actually accept and use in proofs, so that in a sense logicians were being dictated to by mathematicians. But the logicians' analysis of consequence in terms of form is clearly, if implicitly, endorsed by mathematicians such as Hilbert, who in connection with his axiomatization of geometry is famously reported as holding that one must be able to say at all times—instead of planes, lines, and points—"tables, chairs, and beer mugs."[10] Hilbert's assertion amounts to a concise endorsement of the view that a genuinely rigorous deduction of a conclusion from his postulates for geometry must lead to a result that would be equally true of any nongeometric entities and relations that happen to satisfy the conditions his postulates say are satisfied by such entities as *points* and such relations as *lying between*.

Such a view is already to some degree implicit in the use of models, which is to say, of reinterpretations of "plane" and "line" and so on, in connection with non-Euclidean geometry, and also in discussions of the principle of duality in projective geometry. Hilbert exploits the method of reinterpretation systematically to show that his most important axioms are irredundant and cannot be dropped: For any given one of them, there are models, or reinterpretations of "point" and "line" and so on, where *it* fails, though the other axioms hold.

Implicit in all this is what we may call the *paradox of rigor*, namely, the observation that *a treatment of a given subject matter that is genuinely rigorous will ipso facto cease to be a treatment of that subject matter (alone)*. It will always be equally a treatment of any other subject matter where conditions alike in logical form to the postulates in question are satisfied. This clearly implies what we have already seen, that a genuinely rigorous treatment of geometry cannot rely on spatiotemporal intuition.

A more precise statement would be that it cannot rely on spatiotemporal intuition *for the justification of theorems*. Intuition may be, and indeed according to Hilbert definitely is, what suggests the postulates in the first place. Moreover, intuition may aid in the *discovery* of theorems, even if one is not allowed to appeal to it in their proofs, and insofar as it may aid in the discovery of lemmas, little theorems on the way to big

---

[10] Reportedly said to Otto Blumenthal, to whom the Hilbert biography Reid 1970 is dedicated, and used as a chapter title in that work.

theorems, it may also play a role in the discovery of proofs of big theorems, though appeal to intuition does not appear in the final enunciation of such proofs.

These last points are ones that Hilbert, though perfectly aware of them, does not perhaps sufficiently emphasize, leading to a misimpression in some circles that he was somehow opposed to intuition, whatever that would mean. That conclusion is, however, historically absurd. To mention just one reason, Hilbert was the author not only of the rigorous *Foundations of Geometry* but also of the semipopular Hilbert and Cohn-Vossen 1952, whose original German title *Anschauliche Geometrie* means precisely *Intuitive Geometry*. No one, in fact, was better aware than Hilbert of the indispensability of intuition *in its proper place*. It is just that for Hilbert, as for other rigorists, that place is not in proofs.

To go with the paradox of rigor, there is another paradox pertaining to "foundations," the keyword in Hilbert's title. An old simile—I do not myself know how far it can be traced back—compares further and further rigorous deductive development from given postulates to the construction of higher and higher stories in a building, with the postulates at the beginning in the role of the foundation at the bottom. In a real building, the foundation supports the upper stories, and so the metaphor is really only appropriate if one thinks of the postulates as "self-evident," or anyhow, as more evident than what is deduced from them. But just this "foundationalist" point of view is what the modern rigorization of mathematics had to give up.

Though one still speaks of "foundations" today, often all that is really meant is "starting point" or "framework." The choice of postulates (and primitives) is as often as not justified not by their immediate evidence, but by the verifiable correctness, or at least the plausibility of the results derived from them. Moreover, with the wisdom of hindsight one can see that "foundationalism" almost inevitably *had to* be given up, given the need for a commodious common framework for *all* the various branches of mathematics, traditional and modern. A framework sufficient to accommodate all of that, including the most recent and daring extrapolations beyond tradition, will almost inevitably have to be *less* and not *more* evident than the basic principles of the oldest traditional branches of mathematics, such as pure number theory.

This insight may be expressed in a *paradox of foundations*, namely, the observation that *anything that is sufficient as a foundation for all of mathematics (including all of its newest branches), will ipso facto fail to be a foundation for some of mathematics (including some of its oldest branches)*. In this paradoxical formulation, "foundation" is used equivocally, being taken in the loose sense of *framework* in the first clause, and in the strict sense of *support* in the second clause. What the paradox points to is the fact that there is almost bound to be some kind of trade-off between power or comprehensiveness, on the one hand, and intrinsic evidence or certitude, on the other hand. The investigation of the options for such trade-offs is the core topic of mathematical logic today, insofar as it is concerned with "foundational" issues, though a more precise statement is not possible at this stage.

Set Theory

The direction from which a commodious common framework for mathematics eventually emerged was a quite unexpected one, one the rigorists of the middle nineteenth century could hardly have anticipated. For that framework turned out to be provided by *set theory*, a daring innovation which its founder, Cantor, originally introduced (like dynamical systems theory, algebraic topology, and other new branches of mathematics dating from the late nineteenth or early twentieth century) in a form less than ideally rigorous. The basic notions and notations of set theory—for instance, the distinction between the elementhood or membership relation $\in$ and the subset or inclusion relation $\subseteq$, and so forth—are today taught even in primary school, but the subject originated in a quite advanced theoretical context.

For set theory arose from some of the generalizing and rigorizing developments just surveyed, and in the first place from the drive to extend the class of functions considered in analysis, and the drive to rigorize the resulting more general field. The particular form of generalization involved concerned the representation of functions not by power series

$$f(x) = a_0 + a_1 x + a_2 x^2 + a_3 x^3 + \cdots$$

such as we considered earlier for the sine and cosine function, but by *trigonometric series*, involving the sine and cosine functions themselves, thus:

$$f(x) = c + a_1 \cos x + b_1 \sin x + a_2 \cos 2x + b_2 \sin 2x + \cdots$$

Such series arise naturally in the representation of sound waves as combinations of a fundamental tone and overtones, and they appeared already in the eighteenth-century study of the vibrating string problem, but a new impetus was given to the development of the theory of such series in the early nineteenth century by their use by Joseph Fourier in his theory of heat.

Fourier's work was (to put it mildly) not especially rigorous, and a number of results were stated as if universally valid that are only generically valid. Determination of the exact scope and limits of various theorems was a substantial project. The technical details of the problem on which Cantor was working need not concern us. Suffice it to say that he found that a certain result held for a given function provided it was "well-behaved" in a certain sense at every point, and that indeed the theorem still holds if there are some points where the function is not well behaved, provided there are only finitely many such exceptional points.

Indeed, he found that one could allow infinitely many exceptional points, provided they are all *isolated* from each other, meaning that for each exceptional point x there are a and b with $a < x < b$ such that there are no other exceptional points besides x in the interval between a and b. Indeed, one can allow finitely many doubly exceptional points, or exceptional points *not* isolated from all other exceptional points in the sense just explained. Indeed, one can allow infinitely many doubly exceptional points, provided they are all isolated from each other. Indeed, one can allow finitely many triply exceptional points. And so on.

The first conceptual innovation in Cantor's work on this topic was to switch from speaking in the plural of "exceptional points" to speaking in the singular of "the set of exceptional points." He defined a *set* as

any collection into a whole M of definite and separate objects m of our intuition or our thought ... the *elements* of the set. (Cantor 1915, 85)

This definition has been compared to Euclid's definition of point, but it does at least insinuate certain features sets are supposed to have: A set is determined by its elements, so that we cannot have two distinct sets with exactly the same elements, and inversely the elements are determined by the set, so that *these* elements and *those* other elements can never constitute one and the same set. Likewise, the elements must have in some obscure sense a kind of priority over the set, since one can only collect together into a whole what are there to be collected. The elements come first and the set is formed by gathering them, and it is not the case that the set comes first and the elements are formed by dividing it up. But since it is not a question of literal, physical gathering, it cannot be a question of literal, temporal "priority."

What the reification or hypostasis of sets really amounts to is just this, that the set can be treated as a single object *on which operations can be performed*. In Cantor's earliest work, concerned with sets of real numbers or points on the line, the most important operation was that of taking the set-derivative, the operation that applied to a set E gives us the set ∂E of *non-isolated* points in E. Define $\partial^0 E$ to be E itself, and $\partial^{n+1}E$ to be $\partial(\partial^n E)$, so that for instance $\partial^1 E = \partial E$, $\partial^2 E = \partial(\partial E)$, and so on. In these terms, Cantor's result was that his theorem holds if for some n the nth derivative $\partial^n E$ is finite. This includes the case where it has just one element or no elements, giving a reason to posit as ideal elements *singleton* sets with just one element and the *empty* set with zero elements, despite the use of the plural "definite, well-distinguished objects" in his definition of set.

More than this, Cantor showed the theorem still holds even if all $\partial^n E$ are infinite, provided the *intersection* $\cap_n \partial^n E$ or set of points belonging to *all* the various $\partial^n E$ is finite. Cantor defined this to be the "infinitieth" derivative $\partial^\infty E$, and showed that even it can be allowed to be infinite provided the "infinity-plus-first" derivative $\partial^{\infty+1}E = \partial(\partial^\infty E)$ is finite. And so on. Eventually, he substituted lowercase omega, the last letter of the Greek alphabet, for the infinity symbol, in counting the number of times an operation is performed beyond any finite number of times introduced new beyond-the-finite numbers. In grammar, the *cardinal* numerals are *one, two, three,* and so on, while the *ordinal* numerals are *first, second, third,* and so on. Cantor spoke of the *transfinite ordinal numbers*

$$\omega, \omega + 1, \omega + 2, \ldots, \omega + \omega = \omega \cdot 2,$$

$$3 \cdot \omega, 4 \cdot \omega, 5 \cdot \omega, \ldots, \omega \cdot \omega = \omega^2$$

$$\omega^3, \omega^4, \omega^5, \ldots, \omega^\omega$$

and so on. As the notation suggests, he defined sum and product and power operations on his new numbers, thus introducing a *transfinite ordinal arithmetic*. This was his second conceptual innovation.

Other early work of Cantor was remotely derived from ancient questions about straightedge-and-compass constructions. A real or complex number is called *algebraic* if it is a solution to a polynomial equation

$$a_n x^n + a_{n-1}x^{n-1} + \cdots + a_2 x^2 + a_1 x + a_0 = 0$$

wherein the coefficients a_i are rational numbers, and otherwise is called *transcendental*. Permutation arguments subsumable into Galois theory reduce the problem of proving the impossibility of squaring the circle to proving the transcendence of the real number π. It is not obvious, however, that there are *any* transcendental numbers.

Joseph Liouville first proved their existence using approximation arguments. One must first define, as Heron does not, what it is for a rational number to be a "good" approximation to an irrational number. Generally speaking, a fraction m/n in lowest terms is a good approximation to x if the absolute difference $|x - m/n|$ is "small" relative to $1/n$, and the smaller the better.[11] Liouville proved there are limitations to how well an algebraic irrational number can be approximated by rational numbers, and showed that there are irrational numbers that can escape these limitations and be approximated better than any algebraic number can be; such numbers must therefore be transcendental. Building on his work, Charles Hermite proved the transcendence of e, the base of the natural logarithms, and building on *his* work, Ferdinand von Lindemann proved the transcendence of π.

Cantor's work (dating from about the same time as Hermite's, and before Lindemann's) shows that transcendental numbers are the rule, rather than the exception: There are more transcendental than algebraic numbers. Talk of "more" here presupposes some account of comparison of "how many" for infinite sets, some notion of *transfinite cardinal number*. This was Cantor's third conceptual innovation. He takes the existence of a one-to-one correspondence between the elements of two sets as the criterion for sameness of number of elements, or *equinumerosity*, and accepts the consequence that Galileo had found paradoxical, namely, that there may be violations of the "axiom" that the whole is greater than the part, in the sense that a proper subset of a given set may turn out to have the same number of elements as the given set. That, indeed, is the key to his work on transcendental numbers.

He defines an infinite set to be *countably* or *denumerably* infinite if it can be put into one-to-one correspondence with the positive integers or whole numbers, and then proceeds to give a number of examples of sets of which the positive integers are a proper part or subset, but which nonetheless are countable, and so have no more elements, by his criterion, than the set of

---

[11] See Rademacher and Toeplitz 1966, ch. 17.

positive integers. Such to begin with are the nonnegative integers or natural numbers, where the correspondence is the obvious one:

$$1 \quad 2 \quad 3 \quad 4 \quad \ldots$$
$$0 \quad 1 \quad 2 \quad 3 \quad \ldots$$

Similarly the integers (including negatives) are countable, and more surprisingly the rational numbers.[12] So, too, finally are the algebraic numbers. This last result can be construed as a special case of the result that the number of finite expressions that can be formed from a finite alphabet or keyboard of symbols is always countable (since each algebraic number can be described by such an expression, as the such-and-suchth solution in some convenient ordering of such-and-such a polynomial equation, there being only finitely many solutions to any one such equation). That fact is familiar today from the way messages are transmitted electronically, by conversion into a sequence of zeros and ones, the digits of some binary numeral representing what may be thought of as a code number for the expression.

On the other hand, Cantor shows that the real numbers or points on the line are equinumerous with the complex numbers or points in the plane, a result related to Peano's later space-filling curve, and are also equinumerous with the subsets of the set of positive integers. However, there are more real numbers than positive integers, hence more real or complex numbers than algebraic numbers; in other words, as asserted at the outset, transcendental numbers outnumber algebraic numbers. Cantor gives the name $\aleph_0$ to the number of positive integers (or algebraic numbers) and c (usually written in Fraktur) to the number of real (or complex) numbers, which he calls the *power of the continuum*. These are his first two examples of transfinite cardinal numbers.

As with ordinals, so with cardinals, he introduces an arithmetic. The facts about the number of naturals, integers, rationals just mentioned are exemplifications of more general facts expressed arithmetically by saying

$$\aleph_0 + 1 = \aleph_0 \quad \aleph_0 + \aleph_0 = \aleph_0 \quad \aleph_0 \cdot \aleph_0 = \aleph_0$$

---

[12] The proofs of this and other elementary results of set theory to be mentioned are too well known to bear repetition here, but in case any reader should be unfamiliar with them, there are popular accounts in Rademacher and Toeplitz 1966, ch. 7, and in Hahn 1956b.

The fact about the number of reals just mentioned is the instance for $\kappa = \aleph_0$ or the general fact that $\kappa < 2^\kappa$, a compact way of saying that *any* set has more subsets than elements. This immediately gives us a third transfinite cardinal beyond $\aleph_0$ and c, namely, 2^c, after which further exponentiation gives us still more. Note that these equations make it difficult to recognize such "numbers" as $-\aleph_0$ or $1/\aleph_0$, and readers will not be surprised to be told that Cantor was a militant critic of infinitesimals (though his motives seem, post-Robinson, to be confused).

Though Cantor did not much distinguish the finite ordinals and finite cardinals from each other and from the whole or natural numbers, his work does underscore the fact that even the positive integers have two aspects, ordinal and cardinal (which generalize in different ways to the transfinite). For instance, for the multiplication of two positive integers m and n we have distinguishable, though equivalent, ordinal and cardinal characterizations. On the one hand, we may obtain the product $m \cdot n$ by starting with zero and repeatedly adding m, counting the number of times we do so, until we have done it n times. On the other hand, we may obtain the same product by considering an m-by-n array, which is to say, by considering how many ordered pairs (a, b) may be formed with the first element a coming from some set A of size m and the second element b coming from some set B of size n. It is typical of nineteenth-century generalizations of traditional spaces and number systems that they cast new light on familiar matters in this way.

Further examples of countable sets are provided by the ordinals mentioned earlier, through ω^ω and beyond. Each of these is countable in the sense that the set of smaller ordinals is countable. This can in each case be shown by rearranging the positive integers to get a one-to-one correspondence. For the ordinals less than $\omega + \omega$, one correspondence looks like this:

1	2	3	...	ω	$\omega + 1$	$\omega + 2$	...
1	3	5	...	2	4	6	...

The rearrangement simply places the odd numbers first and the even numbers after. A further example of an uncountable set was provided by the set of all countable ordinals. The number of these Cantor called $\aleph_1$. Actually, Cantor did not believe it was a *further* example of an uncountable cardinal, since he conjectured that $c = \aleph_1$, a statement known as the *continuum hypothesis* (CH). Cantor could not prove it, and Hilbert placed it first on his famous list of 23 problems for mathematicians of the then-dawning twentieth century.

Obviously, Cantor's way of thinking goes very much against the nineteenth-century mainstream idea of eliminating "actual infinity," which we have seen was closely bound up with Weierstrassian strategies of rigorization for the calculus and analysis. As for the rigor of Cantor's own work, for the most part we can say what Oxtoby said of Poincaré, that "read against a background" of later ideas, most of his arguments are more or less rigorous as they stand, and there is certainly no error in attributing many basic results of set theory to its founder. Nonetheless, there were in Cantor's work some lapses of rigor that by hindsight are quite glaring. These may be summed up under two heads: *unconscious uses of the axiom of choice* and *failure to state his existence assumptions explicitly* (except to some extent in private correspondence). Let me take these both up in turn.

An example of the first deficiency occurs in Cantor's "proof" that $\aleph_0$ is the *smallest* transfinite cardinal, which is to say that every infinite set I contains a subset consisting of distinct elements $a_1, a_2, a_3, \ldots$ in one-to-one correspondence with the positive integers. Basically, Cantor's procedure is simply to say that since I is infinite and therefore nonempty, it must contain some element a_1, that since I is infinite and therefore has more than just one element, it must contain some other element a_2, that since I is infinite and therefore has more than just two elements, it must contain some yet other element a_3, and so on. It takes some thought to see that there is an unstated assumption being invoked here.

A rigorous proof today would acknowledge that it is being assumed that if we have a set of nonempty sets, such as the set of *cofinite* subsets of I (the sets obtainable from I by omitting finitely many elements), then there is a function f assigning each set J in the given set of sets one of its elements. We then let $a_1 = f(I)$, let I_1 be I minus a_1, let $a_2 = f(I_1)$, let I_2 be I_1 minus a_2, let $a_3 = f(I_2)$, and so on. The assumption that there always exists such a "choice function" as f is known as the *axiom of choice*. It also has several equivalent formulations. The case indicated is not the only one in which Cantor and other pioneering set theorists implicitly assumed the axiom without explicitly stating it as a postulate, as was later done by Ernst Zermelo.

More important is a second deficiency. Cantor's definition of set insinuates the postulate of *extensionality*, according to which sets with exactly the same elements are exactly the same set. This means that *if* there is a set whose elements comprise all and only those x satisfying some condition

$\varphi(x)$, then there is a *unique* such set. The modern notation for *the* set of all and only those x such that $\varphi(x)$ is just $\{x: \varphi(x)\}$. What Cantor's definition does not indicate is any answer to the question *for which conditions $\varphi(x)$ does the set $\{x: \varphi(x)\}$ exist?*

Cantor contrasts "inconsistent multiplicities" or pluralities that *cannot* be collected together into a whole with "consistent multiplicities" or pluralities that *can* be collected together into a whole, and virtually identifies that distinction with the distinction between "absolutely infinite" and "merely transfinite" pluralities, multiplicities *too many* to be collected together and multiplicities *few enough* to be collected together. But, except to some extent in private correspondence, he states no explicit postulates about how many are too many, and how many are not. The development of his thought in this regard is traced in detail in Hallett 1982, but it never issued in a clear public enunciation of postulates.

Paradox

In the absence of any clear guidelines from Cantor, some later writers fell into paradoxes. The first published set-theoretic paradox, due to Cesare Burali-Forti, was highly technical, and the last to be discovered, due to Julius König, involves the problematic notion of "definability." But in between there came a basic paradox, making only quite immediate, non-technical use of basic, pure set-theoretic concepts, that arises as soon as one assumes, unlike Cantor, that there exists a set of all entities x, or even just a set of all sets x. For with such a "universal set" V, every subset will already be an element, and this contradicts Cantor's theorem that the subsets always outnumber the elements.

By analyzing Cantor's proof of his theorem as it applies to this case, Russell was led to consider the set of all entities, or of all sets, that are not elements of themselves, the so-called *Russell set* $R = \{x: x \notin x\}$. If one asks whether it is an element of itself, whether $R \in R$, we more or less immediately reach the contradictory conclusion that it is if and only if it isn't. Russell found the paradox while considering the system of Frege, who hoped to reduce all of arithmetic, algebra, and analysis to "logic" in a broad sense. Frege, unlike Cantor, did explicitly assume that *every* condition determines a set or class, so his system was straightforwardly inconsistent. (The trouble with Cantor's work, by contrast, was not inconsistency in its basic assumptions, but rather lack of definiteness about what its basic assumptions *were*.) The "Russell paradox" was apparently independently discovered by Zermelo about the same time.

Russell and Zermelo offered interestingly different responses to the paradox. Russell, who still wished to carry on with the "logicist" project of reducing all of mathematics to pure logic, proposed a revision of Frege's system called the *theory of types*. Jointly with Whitehead, he developed a substantial amount of mathematics within the framework of this system in the three-volume *Principia Mathematica*. (A fourth volume was projected, but the authors became exhausted.) The system was baroque in its complexity, and the theory of types as studied today—with its offshoot, the so-called *typed λ-calculus*, it is prominent in certain areas of theoretical computer science—generally derives from a simplification by Russell's student F. P. Ramsey, with later refinements. Zermelo, by contrast, proposed an *axiomatic set theory*, with explicit existence assumptions of the kind that Cantor conspicuously failed to publish. His proposal, with certain modifications, was the one that achieved widespread acceptance. (Actually, Ramsey's simplified theory of types can also be viewed as a system of axiomatic set theory, a little weaker than but ultimately not all that different from Zermelo's.)

According to the detailed study Moore 1982, Zermelo proposed his axiom system less on account of the paradoxes than as a setting for his work on the axiom of choice, and so it will be in order to say something about the latter. The natural numbers with their usual order are *well* ordered, in the sense that every nonempty subset has a least element. The integers, rationals, and reals in their usual ordering are *not* well ordered. For instance, the set of *positive* rationals or reals has no least element. The proof of the countability of the rationals shows, however, that they can be *rearranged* so as to be well ordered, and indeed ordered just like the positive integers. Cantor raised the question whether the set of real numbers, which by contrast with the rational numbers are *un*countable, could also be *rearranged* in some way so that it would be well ordered.

This Zermelo proved (despite a supposed counterproof by König, involving his definability paradox), but only assuming the axiom of choice (AC), in a formulation he attributed to the analyst Erhard Schmidt. The axiom at once became controversial, but let me postpone saying anything about that controversy. For the moment, suffice it to say that Zermelo's paper on the well-ordering theorem was fairly quickly followed by another on the axiomatization of set theory, in which AC appears as merely one of several postulates, alongside the principle of extensionality, already discussed, and a number of specific existence assumptions.

Instead of assuming that for every property P there is a set of all and only those x having the property—an assumption that leads to the Russell paradox—Zermelo assumed only that, given any set a, we can separate out from it the set of those of its elements that have the property P. Thus he assumes in his *axiom of separation* the existence for every a of the set

$$\{x \in a: P(x)\} = \{x: x \in a \text{ and } P(x)\}$$

Of course, this much does not give us any new sets unless we already have at least one old set a to get going with. Such sets were provided by his other existence axioms. These provide for the existence of the *empty* or *null* set, the *unordered pair* of any two elements, the *union* of a set of sets, and the *power set* of any given set:

$$\emptyset = \{x: x \neq x\}$$

$$\{a,b\} = \{x: x = a \text{ or } x = b\}$$

$$\cup a = \{x: \text{for some } y, y \in a \text{ and } x \in y\}$$

$$\wp(a) = \{x : x \subseteq a\}$$

A crucial fact, and one that led to set theory eventually taking center stage in the program of rigorization, is that *these are essentially the existence assumptions needed to get new spaces or number systems or whatever from old ones by forming products and powers and quotients and so forth*, in the manner of nineteenth-century introduction of auxiliaries for the study of traditional spaces or number systems. Indeed, the constructions of the auxiliaries in question can be, and now in retrospect are, viewed as essentially "set-theoretic" constructions, though some of them actually antedate Cantor.

All the postulates mentioned so far do not, however, suffice to prove the existence of infinite sets. To get going, Zermelo further assumes the existence of an infinite set, like the set of natural numbers. With this additional assumption one can reconstruct all the traditional spaces and number systems in a set-theoretic context, and we thus have a framework adequate to encompass all of mathematics, despite the vast expansion that took place in the nineteenth century.

Or anyhow, we almost do. Some additions and amendments may be called for. For one thing, Adolf (later Abraham) Fraenkel and Thoralf

Skolem were both dissatisfied with one feature of Zermelo's axiomatization, his dependence on an axiom of separation formulated in terms of a notion of "property." Their criticism led to an amendment in which that "second order" single *axiom*:

> for any property and for any set *u* we can form the subset of *u* whose elements are the elements of *u* that have the property

involving a notion of "property" left without formal characterization, was replaced by a "first-order" axiom *scheme*, or rule to the effect that any statement of a certain general form is to be counted as an axiom, namely, the rule for each condition $\varphi(x)$ that can be built up logically (using negation, conjunction, disjunction, and universal or existential quantification over sets) from the primitive notions "$x = y$" and "$x \in y$", the following is taken to be an axiom:

> for any set *u* we can form the subset of *u* whose elements are the elements x of u such that $\varphi(x)$.

For another thing, just as Cantor overlooked the need for an axiom of choice, Zermelo overlooked the need for an axiom of *replacement*. According to this axiom (nowadays understood as a scheme, like separation), if to each x we have associated a unique y—call it y_x—satisfying some condition $\psi(x, y)$, then we can take any set a and replace each x in a by the associated y_x to obtain the set

$$\{y_x : x \in a\} = \{y : \text{for some } x, x \in a \text{ and } \psi(x, y)\}$$

Adding this axiom scheme to the amended version of Zermelo's system produces *Zermelo-Fraenkel set theory with choice* (ZFC). That system, or some simple variant—the best known is the system NBG of John von Neumann, Paul Bernays, and Kurt Gödel—is the one enjoying the widest acceptance today.

As some of my remarks have already intimated, it suffices as a "foundation"—or less misleadingly and pretentiously, a "framework"—for traditional mathematics plus its many nineteenth- and twentieth-century extensions by the introduction of ideal elements and other auxiliaries. The situation by about the middle of the twentieth century has recently been described by the logician Wilfrid Hodges in the following terms:

> [D]evelopments led to a new a picture of the logical structure of mathematics. The picture is not associated with anybody's name in particular; it just happened . . .

[F]rom the 1950s onward, classical mathematics had just one deductive system, namely, first-order Zermelo-Fraenkel Set Theory with [the Axiom of] Choice . . . (Hodges 2008, 473)

Hodges adds that the traditional "axioms" of, for instance, group theory, came to be regarded simply as the definitions of certain kinds of spaces or number systems or whatever within the universe of set theory. What Hodges describes is essentially the situation today.

So long as AC remained somewhat controversial, theorems depending on it were sometimes starred, like sports records set by athletes on steroids. Such starring of choice-dependent results has lapsed, but anything depending on assumptions beyond ZFC would still today be literally or metaphorically starred. Assumptions going beyond ZFC used in proofs must be acknowledged as hypotheses in the theorems proved; assumptions included in ZFC do not require special mention. Such is the policy of the prestigious *Annals of Mathematics*, for instance.[13]

Anti-Rigorism

So, the nineteenth-century project of rigorization extended well into the twentieth century, and culminated in the development of mathematics within the framework of an axiomatic set theory like ZFC. I have left the question why mathematicians first undertook the project of rigorization that led to this result, one entirely unanticipated by the initiators of the program, in a rather unsatisfactory state, pleading that motives were many and mixed. A related and perhaps more promising question would be a slightly different one: Why, once the rigorized version of mathematics was finally arrived at, did it win general acceptance? I mean, of course, acceptance in the community of mathematicians, since as I have said, no one expects the same degree of rigor of physicists, let alone engineers. Their disciplines and professions have their own priorities.

This new question of acceptance is perhaps best addressed by considering the standpoint of *opponents* of the program of rigorization, and why they in the end failed. Along with opposition to rigor, we must consider opposition to set theory, since of the alternatives available only axiomatic set theory somewhere in the neighborhood of ZFC offered a prospect for a

---

[13]   Or so I have it third-hand. A slight qualification will appear later on in this study.

rigorization commodious enough to take in all the new branches of mathematics that had sprouted since 1800. But it should be kept in mind that this inextricable intertwining of rigorist and set-theoretic ideas was not apparent to everyone at the time, so that circa 1900 one could very well identify oneself as a proponent of rigor and opponent of set theory. Of all the many influential oppositionists, there will be space to consider just three: Oliver Heaviside, plus Hilbert's archenemy Brouwer, and a figure already repeatedly mentioned, Poincaré. The first was an outright anti-rigorist, while the other two opposed the emerging set-theoretic orthodoxy.

To begin with Heaviside, there is no need to discuss here his contributions to developing the vector-calculus notation, still very much in use, or his "operator methods" for solving ordinary differential equations, still employed as a heuristic device. More significant in the present context is his work on applications of divergent series. For he took himself, in demonstrating their utility, to be scoring a major point against those "wet blankets," the rigorists.

But rigorists need not regard divergent series the way Pythagoreans regarded beans, as something absolutely to be eschewed in all circumstances. It is a requirement of rigor that one must seek to determine precisely the scope and limits of methods that work only "generically" and not "universally," and that in doing so one must give rigorous definitions of all notions involved. It is no requirement of rigor that one must stop with the first rigorous definition that one finds in a given area, usually the simplest and most natural one. Rigorism imposes no restriction on definitions other than rigor, and leaves one perfectly free to explore definitions that may at first seem complicated and unnatural, if there are grounds for hoping they will be useful and fruitful. Convergence may be the simplest and most natural rigorously defined notion for use in connection with series, but there is no ban on or bar to exploring other definitions, if there are useful generic-but-not-universal phenomena that cannot be explained in terms of convergence considerations alone.

To illustrate this point in the case of Heaviside's own use of "asymptotic" series would take us too far into technicalities, but the point can be illustrated in a less technical way in a related case.[14] The simplest and most

---

[14] Kline has a special liking for the topic of divergent series and devotes his ch. 47 to this rather restricted topic. That chapter won't all be easy reading for the non-mathematician, but it is worth looking at even by the reader who doesn't want to go deeply into the technicalities of the subject.

natural definition of "sum" for an infinite series is the one applicable to convergent series. But we are free to devise other definitions if we can. One route to an alternative begins with the observation that if the sequence s_1, $s_2, s_3, \ldots$ of partial sums converges, then the sequence of averages

$$t_1 = s_1$$
$$t_2 = (s_1 + s_2)/2$$
$$t_3 = (s_1 + s_2 + s_3)/3$$

also converges, and to the same limit. But it may happen that the t_i converge even though the s_i do not. Indeed, this happens with the alternating series, where the s_i and t_i are as follows:

1	0	1	0	1	...
1	1/2	2/3	1/2	3/5	...

Nothing prevents us from defining a notion of "summability" and which the alternating series counts as summing to $1/2$ on the strength of the sequence of averages converging to that limit, even though the series does not converge in the sense that the sequence of partial sums converges. In so doing, we are merely introducing a new rigorously defined notion that subsumes and generalizes an older one (and one that just happens to give sense to the claim that the "sum" of $1 - 1 + 1 - 1 + \cdots$ is $1/2$).

Owing to the freedom to introduce useful new notions regardless of simplicity or naturalness, so long only as they are rigorously defined, *nothing need be lost*. Some extra work is required, but at the cost of this work we achieve the benefit of replacing vague "generic-but-not-universal" results with results whose scope and limits are precise. To the extent that this is so, Heaviside's complaints about "wet blankets" are not well taken. In any case, there is the division of labor to be considered: Restrictions of rigor are only meant to apply to mathematicians, not physicists, let alone engineers; and despite his making more contributions to mathematics than many respectable mathematicians, Heaviside was an engineer.

Time and again during the early twentieth century non-rigorous but useful methods introduced by physicists or engineers were found to have rigorous counterparts, the development of which at the very least clarified the scope and limits of the techniques, and more often than not had other side benefits as well. A much-cited case is that of the δ-"function" of

Paul Dirac, whose rigorization led to the *theory of distributions* of Laurent Schwartz.

In the face of such successes, Heaviside's type of anti-rigorism simply need not be taken seriously, at least not nowadays. To be sure, there are still methods being used by physicists, especially in quantum field theory, whose rigorous basis is not well developed or well understood. But in such cases, while no physicist would agree to suspend use of such methods pending mathematicians' discovery of a way to rigorize them, many would agree that the situation is not one with which one can be permanently satisfied. To the extent that methods *persistently* resist rigorization, one has reason to suspect that they will eventually be superseded. One can say something like this because by now it has turned out time and time again that methods that are really physically useful ultimately can be put on a basis that is mathematically rigorous.

There remains, of course, the question whether the benefit of doing things rigorously, beginning with the benefit of getting a clear idea of when certain methods are safe to apply and when not, is worth the cost in extra work. And here Borges's Pierre Menard phenomenon, mentioned in connection with the Poincaré recurrence theorem, becomes relevant again. Often hardly any extra work is needed, once certain fundamentals have been cleared up. More generally, doing things rigorously will seem much less work to those who have been brought up since their student years to do things that way, than to those used to a freer and sloppier way of working, who suddenly face the "wet blanket" of new requirements of rigor. What one has here is simply an instance of a general feature of scientific revolutions noted by many commentators, that they often triumph simply by the older generation dying off and a new generation coming to the fore.

Intuitionism and Constructivism

While the English engineer Heaviside thought mathematicians' standards of proof were becoming too strict, the famous Dutch topologist Brouwer held that in an important sense they were too loose, since they allowed *nonconstructive existence proofs*, purported proofs that there exists a mathematical object with some mathematical property that do not even implicitly offer a method for identifying a specific example. (On these grounds, he held that even his own most famous result, the *Brouwer fixed point theorem*, requires a "correction.") While Heaviside merely complained about the direction in which mathematics was moving, Brouwer campaigned to change the direction of motion, founding a dissident movement known as *intuitionism*.

Unease over nonconstructive existence proofs was fairly widespread, and Brouwer succeeded in attracting at least one very prominent mathematician of the rising generation, Weyl, to his movement. For a time it seemed to Hilbert that there was a serious possibility that many more might be drawn away from his own style of mathematics. And yet Brouwer's movement failed—failed not only to deflect mainstream mathematics, but even to establish any very powerful sidestream. Intuitionism could not even permanently retain the allegiance of Weyl.

Many reasons of varying degrees of importance may be cited for Brouwer's failure to attract more than a small group of disciples. On the one hand, Hilbert used some strong-arm tactics in what Einstein called a "frog and mouse battle" in academic politics, managing, for instance, to eject Brouwer from his powerful position on the editorial board of the *Mathematische Annalen* at a time when it was still the leading journal in the mathematical world. On the other hand (though this aspect of his thought may not at first have been widely known outside the Netherlands), Brouwer mixed up his distinctive approach to mathematics with a mysticism hard-headed scientists were bound to find rebarbative.[15]

Moreover, Brouwer's mysticism led him to believe that mathematics is largely an *incommunicable* mental activity, not to say an *ineffable* meditative practice, and as a result the rules of intuitionist game were never clearly articulated. No wonder dissentions arose even among his close disciples.[16]

---

[15] Benacerraf and Putnam 1983 includes and makes conveniently available key selections from most modern philosophers of mathematics whom I will have occasion to mention: Brouwer, Dummett, Frege, Hilbert, Quine, Russell, as well as from the editors themselves. But readers should be warned that the anthology contains only an expurgated version of Brouwer's "Consciousness, Philosophy, Mathematics" (90–6), keeping the Kant but dropping the Krishna. Something more radical was attempted a half-century afterwards by the late Michael Dummett (in Dummett 1975, among other places), where Brouwer's own motivation for intuitionism, which always tended towards solipsism, is replaced by a radically different one, tending towards behaviorism. Whatever the merits of this approach, it came far too late, long after the mathematical community had made up its mind. Oxford may no longer have been crawling with Jacobites toasting "the king over the water," but Matthew Arnold's description of that university as "the home of lost causes" remained applicable in the Dummettian era.

[16] One Dutch mathematician discussing the situation with me many years ago also cited a supposed feature of the national character of his compatriots, reflected in an alleged old proverb "One Dutchman a Christian, two Dutchmen a church, three Dutchmen a schism."

Exactly *what* the alternative to the nonconstructive practice that was rapidly becoming "classical" was supposed to be never became entirely clear. Can one "define" a function by reference to the mental activities of a creative subject? Can one "define" a sequence by saying that its successive terms are to be picked by independent successive acts of free choice? Can one "define" a sequence by considering what one has proved in the course of one's mathematical career and how one has proved it?

There is in any case a general tendency for heterodox movements to become fissiparous, and the general tendency known as *constructivism*—the tendency to question nonconstructive existence proofs—eventually gave rise to a series or spectrum of alternatives to classical mathematics, rather than a single alternative position: Brouwer's *intuitionism*, its offshoot Griss's *negationless mathematics*, the older *finitism* of Kronecker, and several later approaches, including the unsavory Ludwig Bieberbach's nasty *Deutsche Mathematik*, A. A. Markov's *Russian constructivism*, Alexander Yessenin-Volpin's exotic *ultra-intuitionism*, the neo-finitism of Errett Bishop, and the constructive type theory of Per Martin-Löf, this last being by far the most active at present.[17]

Another reason cuts even deeper: the magnitude of the sacrifices that it appeared one would have make for the sake of constructivism. Brouwer himself, in striking contrast to Heaviside, was indifferent to the needs of applications, contempt for the notion of human technological domination of nature being the feature of his mysticism with which it is perhaps easiest for many today to sympathize. But Weyl, for instance, the most prominent early sympathizer with Brouwer, as an important contributor to

---

[17] See Troelstra 2011 for a fairly comprehensive summary. All groups make *negative* use of constructivistic restrictions: In proving an existential statement, all require that the item asserted to exist must be shown to be in some sense constructible. Brouwer is almost alone in also making *positive* use of constructivistic restrictions: In proving a universal statement about all items of some kind, he may use the assumption that any such item can be in some sense constructed. Much of the peculiar flavor of Brouwerian analysis derives from this feature, which does not appear in Martin-Löf constructivism, for instance. The Martin-Löf school in this respect recalls Euclid. Euclid makes no assertions to the effect that a figure F with such-and-such properties exists, or that for any figure E with such-and-such properties a figure F related to it in such-and-such a way exists, except in the form of assertions that a figure F *can be constructed*, or that for any figure E a figure F *can be constructed* from it, where "constructed" means "by straightedge and compass." But when Euclid goes to prove an assertion to the effect that every figure F has some property, he never makes use of any assumption to the effect that any given figure F *must have been constructed* by straightedge and compass. He needs to avoid doing so if the theorems in his early books are to remain applicable, even in later mathematical work where other means of construction become available.

mathematical physics could not neglect the needs of that field, and besides many pure mathematicians were dismayed by how much it appeared necessary to give up for the sake of constructivism.

Many mathematicians were made uneasy by the axiom of choice (AC), which conspicuously affirms the existence of something, a choice function, without giving any idea how to construct one. But long before one comes to AC or any higher set theory—in the 1920s still a comparative novelty many thought they could live without—"classical" mathematics presents us with nonconstructive existence proofs. The stock example is the proof that there exists a pair (a, b) of irrational numbers such that a^b is rational. Using the fact that $\sqrt{2}$ is irrational and that $(\sqrt{2}^{\sqrt{2}})^{\sqrt{2}} = 2$, it is easily seen that $(\sqrt{2}, \sqrt{2})$ is an example if $\sqrt{2}^{\sqrt{2}}$ is rational, while $(\sqrt{2}, \sqrt{2}^{\sqrt{2}})$ is an example if $\sqrt{2}^{\sqrt{2}}$ is irrational. Arguing by cases, which is to say, assuming the *law of the excluded middle*, p-or-not-p, according to which either $\sqrt{2}^{\sqrt{2}}$ is rational or if $\sqrt{2}^{\sqrt{2}}$ is not rational, it follows that an example exists. But this proof does not actually give us an example unless we know whether $\sqrt{2}^{\sqrt{2}}$ is rational or not.[18]

This example is of no importance except as an illustration, but what it illustrates is of great importance indeed: If one is to avoid nonconstructive existence proofs, then one must depart from classical practice at a very early stage. One must eschew the very common practice of arguing by cases or appeal to the law of the excluded middle. Likewise, one must eschew arguing by *reductio*, or what in the end comes to the same thing, appeal to the *law of double negation*, not-not-p-implies-p. These are forms of inference used by mathematicians at least since the time of Euclid. It was the genius of Brouwer to recognize that, if existence proofs are to be always constructive, then one must give them up, and work always in "intuitionistic" rather than "classical" logic, though he left it to his disciple Arend Heyting to articulate just what the rules of such a logic amount to. For many mathematicians, the restriction on logic is simply too great a sacrifice to demand. As Hilbert put it, giving up excluded middle is like depriving a boxer of the use of his fists. It seemed that too much of modern mathematics would have to be given up, essentially *without replacement*, as irreconcilable with intuitionistic restrictions.

And yet one *can* get used to working in intuitionistic logic, to the point that one develops a (non-Kantian, non-spatiotemporal) "intuition" about what

---

[18] It is irrational, but that is a consequence of a deep and difficult theorem of A. O. Gelfand.

is intuitionistically acceptable, and can carry out sustained argumentation without excluded middle or double negation, as a boxer not allowed to punch with his fists may learn to kick with his legs and butt with his head. Moreover, there are definite benefits to be reaped *if* one can manage to prove a theorem intuitionistically and not just classically. For instance, a suitably constructive proof of a theorem of the form "For every positive integer m there is a positive integer n such that . . . " will always implicitly provide a recipe for computing a function g that gives a suitable $n = g(m)$ for any given m, something that is by no means true for all classical proofs.[19] Given the theoretical interest, and in some situations the practical importance, of computation and algorithms, the benefit of getting algorithms as a by-product of "constructive" proofs appears an important one. Furthermore, it has transpired that much more of classical mathematics can be salvaged constructivistically than it originally appeared, the more so if one follows Bishop or especially Martin-Löf rather than Brouwer.[20] But all these positive points came to be appreciated only too late, after the struggle was essentially over.

Another important factor was the development, by Alonzo Church and Alan Turing in the 1930s, of a way of addressing many constructivist concerns *within* classical mathematics. What Church and Turing accomplished was to give a rigorous definition of *computability*. This makes it possible to address the question whether a function from positive integers to positive integers is computable, while recognizing that question as a *distinct* question from the question whether for every m there exists an n. If one is interested in computation, one can address the *computability* question *directly*, while remaining within "classical" mathematics, rather than addressing it *indirectly* by considering whether the *existence*

---

[19] E.g. Lagrange proved that every positive integer can be written as the sum of (at most) four squares (not necessarily all distinct). One Edward Waring conjectured in the late eighteenth century. that for every m there is an n such that every positive integer can be written as the sum of n mth powers, and in the early twentieth century. Hilbert proved as much. Classically, it immediately follows that for every m there is a unique n, call it $g(m)$, such that it is always sufficient and sometimes necessary to use that many mth powers. But no algorithm for computing the function g is known. A sufficiently "constructive" proof would give us such an algorithm.

[20] I sometimes have the suspicion that if Martin-Löf rather than Brouwer had been the opposition leader in the 1920s and 1930s, the outcome of the "foundational struggle" might have been different. The Martin-Löf approach has drawn enough interest that computer "proof assistants" have been developed for it, as for classical mathematics. And it has received a tremendous boost in its public profile from the discovery of an unintended interpretation of its formalism useful in homotopy theory in algebraic topology. What would Weyl have thought of all this?

question can be settled affirmatively in "constructive" mathematics. The further development of computability theory gave rise to a subbranch, *complexity theory*, that allows us to raise yet further questions, such as the following: Is the function *g* computable using an algorithm for which the number of steps of computation required to compute $g(m)$ is bounded by some polynomial function of the number of digits in the numeral for *m*? The existence question, the computability question, the polynomial-time computability question, and more all become rigorously formulated mathematical problems that can be separately or successively investigated. There is no need to assimilate the existence question to one of the others, or to disallow, in investigating such questions, any forms of deductive argument traditionally used by mathematicians.

Predicativism

It remains to say something about the towering figure of Poincaré, who has a reputation as an opponent of set theory, and as no proponent of rigorous deduction. In fact, Poincaré, despite the remark quoted earlier about the Weierstrass function, had no objection of principle to the program of rigorization, including for instance the "degeometrization" or "arithmetization" of analysis. He himself contributed substantially to the rigorous study of "asymptotic" series, which I cited in suggesting how the rigorist might answer Heaviside.

He did oppose the "logicism" of Frege and Russell, the claim that rigorization could be pushed so far as to leave no need for nonlogical postulates at the beginning or the bottom of the chain or tower of rigorous deductions, maintaining instead that one would always need the principle of mathematical induction as a postulate suggested by intuition. But to oppose logicism is not to oppose rigor, whose requirements pertain to how one gets from postulates to theorems, not to where one gets the postulates from; and to reject logicism was hardly a minority position. Hilbert as much as Poincaré saw the need for postulates suggested by intuition rather than logic, and Poincaré as much as Hilbert granted in principle the importance of excluding further appeal to intuition in proofs once the postulates have been set up.

Poincaré's reputation as an opponent of set theory is also exaggerated. It seems largely based on a quotation, repeated from author to author, according to which Poincaré is supposed to have compared set theory to a disease. Gray 1991, however, traces this quotation back to its source and finds it to be a garbled paraphrase of a genuine remark in which Poincaré compared the *paradoxes* to a disease, and set theory to the *patient*. Poincaré himself, Russell, and others were so many attending physicians, differing in their diagnoses and prescriptions, but all agreeing that it was a most interesting case for a pathologist—and none recommending euthanasia.

Poincaré was indeed one of the earliest to adopt and apply some of Cantor's revolutionary ideas. To be sure, he never accepted *all* of Cantor's principles, nor did he accept Zermelo's axiom of choice; and he criticized Zermelo's axiom system on account of the absence of a proof of consistency. His prescription for dealing with the disease of the paradoxes was to impose a severe diet, one that would have left the patient much diminished in size. His remarks pointed in the direction of what has come to be called "predicativism," a philosophy somewhat more liberal than "constructivism," though more conservative than "classicism." Its restrictions pertain less to arithmetic than to analysis, and its requirement is less that functions should be computable than that they should at least be definable (a requirement still very much incompatible with acceptance of AC).

Predicativism found even fewer adherents than intuitionism. The members of the "Paris school" of analysis—including besides Lebesgue the prominent figures of Émile Borel and René Baire—who were sometimes called "semi-intuitionists" on account of their objections to the axiom of choice, might better be called "semi-predicativists"; but the emphasis must be on "semi." Predicativistic restrictions were incorporated into Russell's complicated version of the theory of types, though he introduced an hypothesis ("the axiom of reducibility") that largely cancelled them, and they dropped out completely in the subsequent simplification of the theory. One Leon Chwistek thought they should be insisted upon more seriously, rather than abandoned, but he was a philosopher and a painter, and no mathematician. Weyl embraced predicativism for a time, but gave it up his own attempt and joined Brouwer (for a time). The predicativist view was revived only much later by Solomon Feferman, Kurt Schütte, and other logicians, but logicians seem more concerned to investigate than to advocate the doctrine. A variant more liberal form was advocated by the more militant Paul Lorenzen, who founded a school; but it remained a very small one. Edward Nelson has applied the label "predicativism" to yet another restrictive view, which others have called a form of "strict finitism," insisting that it is really a form of constructivism.[21]

In any case, Poincaré's theorizing raised no serious obstacle to the program of rigorization and set-theorization. The problem was not with his principles, but with his practice, for his own work was pretty persistently non-rigorous, though most especially (and admittedly) so in his last years, when he was eager to present in some form ideas he knew he would not have time to polish. The problem indeed was not really with Poincaré's own personal practice: Genius does as it must, and if this particular genius needed others to follow after him patching up the rigor, well, that has been true of other geniuses before and since. The real problem was with sub-geniuses and nongeniuses following Poincaré's example and adopting in their own practice a relaxed attitude towards rigor.

The result was a younger generation of French mathematicians, influenced by Poincaré, falling behind their German contemporaries, influenced by Hilbert. Or so it seemed, a decade and more after Poincaré's death, to a dissatisfied group among young French mathematicians of the interwar years. Writing under the collective pseudonym "Nicolas Bourbaki," from the 1930s onwards they campaigned to bring Francophone mathematics up and perhaps beyond the level of rigor of Germanophone mathematics. By about 1950, rigorization-plus-set-theory was as complete in France as anywhere else. I will be returning to the Bourbaki group at the beginning of Chapter 3, but for the moment I am done with my sketchy history, and ready to return to the question of the further clarification of what *deduction*, the key ingredient in rigor, amounts to.

Deduction and Deducibility

Here one must confront the fact that, if one looks to logic textbooks, though one will find an agreed characterization of logical *consequence*, which I have already discussed, one finds no agreed characterization of

---

[21] Unfortunately, I know of no survey of the varieties of predicativism comparable to Troelstra's survey of constructivism. I earlier cited, as a partial answer to the question why, given the interest and importance of computability, a larger minority of mathematicians did not adhere to constructivism, the circumstance that Church and Turing managed to develop a theory of computability *within* classical mathematics. To the similar question why, given the interest and importance of definability, a larger minority of mathematicians did not adhere to predicativism, one might cite Alfred Tarski's development of a theory of definability *within* classical mathematics as part of the answer.

logical *deduction*. One finds in different books several different formats for deductions: the "sequent calculus" approach of Gerhard Gentzen, the "axiomatic" approach of Frege and Russell and Hilbert, and the "natural deduction" approach also due to Gentzen. Moreover, for each format one will find different books that adopt it, developing it in different ways.

To be sure, in every book the first goal is to prove the *soundness* of whatever deductive system or proof procedure is adopted, which is to say, to prove that if there is a deduction of some conclusion from given premises, then that conclusion is a consequence of those premises. And the second goal is to prove *completeness*, which is to say, to prove conversely that if some conclusion is a consequence of given premises, then there is a deduction of that conclusion from those premises. This means that all the books agree about *deducibility*: Given any two books and any premises and conclusion, there will exist a deduction of the kind favored by the first book if and only if there exists a deduction of the kind favored by the second book, simply because for either book there will be a deduction if and only if the conclusion is a consequence of the premises.

Nonetheless, such agreement about *deducibility* coexists with disagreement about *deduction*. Moreover, *none* of the logic books' approaches to deduction closely matches the practice of mathematicians in giving what they consider rigorous deductive proofs. The "sequent calculus" approach is utterly unlike anything in mathematical practice. The "axiomatic" approach is also dissimilar to what one finds in the literature, because on the axiomatic approach every single sentence appearing in a deduction or proof is *categorically asserted*. This is quite contrary to the usual practice in mathematics, which often involves *hypothetically supposing* something and drawing consequences therefrom.

Two such forms of hypothetical reasoning have already been mentioned. In (one form of) *proof by cases*, one first assumes p and deduces that q follows under that hypothesis, and then assumes instead not-p and deduces that q follows under that hypothesis also, and then finally concludes categorically, under no special hypotheses, that q. In (one form of) *proof by reductio*, one assumes not-q and deduces under that hypothesis first that p, then that not-p, and then finally concludes categorically, under no special hypotheses, that q. These two forms of hypothetical reasoning happen to be intuitionistically unacceptable, but there are many other forms common to classical and to intuitionistic mathematics; and in any case, our interest now is in the classical current and not the intuitionist eddy.

The "natural deduction" approach, very popular with writers of introductory logic textbooks, is designed precisely to allow hypothetical reasoning. However, it will be found that different textbooks start from different lists of basic forms of hypothetical argument. If textbook X takes proof by cases as one of its basic procedures, it is very unlikely to give proof by reductio the same status, while if textbook Y takes proof by reductio as one of its basic procedures, it is very unlikely to do the same for proof by cases. To be sure, if the methods of textbook Y are sound, and the methods of textbook X are complete, then it must be possible to achieve the effect of anything Y allows as a single deductive step by some series of steps allowed by X, and vice versa. But what is a simple, basic move in one book will generally be a complex, derived move (or rather, series of moves) in the other, and vice versa.

In this respect, *all* the logic textbooks differ from the practice in mathematics textbooks and professional journals. The logician is generally concerned to give some *minimal* list of basic procedures to which all others can be reduced, while the mathematician is prepared to use a much more generous and comprehensive list of forms of deductive procedures, without regarding some of these as more fundamental and others as merely derivative. Mathematical deductions generally telescope what for the logician may be a long series of steps into a single, quick move. This is in many ways a more significant difference between logical and mathematical works than the more striking but more superficial fact that the former uses special symbols "~" and "&" and "∨" and "→" and "∀" and "∃," where the latter will write out in words "not" and "and" and "or" and "if" and "all" and "some".

In addition to such telescoping of deductive steps, some parts of the deduction may be omitted altogether, if they are very similar to parts already given, and hence easily supplied by the reader. Such omissions are marked by expressions such as "an exactly similar argument shows . . . " or "the argument given for part (A) applies *mutatis mutandis* for part (B)," and so on. Mathematical works are also readier than logical works to omit explicit citations of earlier results. These will generally be left out in the case of any results that can be expected to be thoroughly familiar already to the readership of the book or periodical involved: "You don't cite Newton every time you take a derivative," as the saying goes.

Mathematical logic is, like mathematical physics or mathematical economics, in the business of modeling an empirical phenomenon.

It differs from mathematical physics or mathematical economics in that the phenomenon being modeled happens to be that of mathematical theorem-proving activity itself. It resembles mathematical physics and mathematical economics in that its models are highly idealized. In particular, as I have been emphasizing, it gives quite a good model of mathematical *deducibility*, but a much poorer model of mathematical *deduction*.

When mathematical logicians prove, for instance, that the set-theoretic hypothesis CH is not deducible from the set-theoretic axioms ZFC—and Paul J. Cohen received the Fields medal for proving this result by his celebrated method of *forcing*—one can be quite confident that, though the result was proved for an idealized model, it will apply to the empirical phenomenon, that one will be wasting one's time if one tries to prove, by ordinary mathematical standards, the continuum hypothesis from the Zermelo-Fraenkel axioms with choice. Attempted proofs of the continuum hypothesis do continue to appear even after the development of forcing, just as attempts to trisect the angle do continue to appear even after the development of Galois theory; in the one case as in the other, the attempts bear all the marks of being the work of cranks.

By contrast, results about the *lengths* of proofs cannot be so readily transferred from the model to the phenomenon. If there is something that counts as a proof by ordinary mathematical standards, logicians are confident that there could in principle be written down something that would count as a proof by ideal logical standards, but the latter may be very much longer indeed than the former, so much so that though writing it down may be possible in principle, it is not feasible in practice.

This brings us to the grain of truth in the formulation I earlier rejected, "A proof is what convinces," and in variants thereof, such as the version attributed to Mark Kac, "A proof is what convinces a reasonable person; a rigorous proof is what convinces [even] an unreasonable person," or the one attributed to Hyman Bass, "A proof is what convinces mathematicians that a formal proof exists" (either of which pair of insightful formulations really deserves an extended commentary of its own). The grain of truth lies in the fact that what convinces depends on the audience, and accordingly what counts as rigor or proof in mathematics depends on who is being addressed.

Mathematical practice comes closest to the logical ideal at the level of textbooks for advanced undergraduate mathematics majors and beginning mathematics graduate students. Below this level, and in works

intended for students not only of mathematics but of science and engineering, it may be thought that the reader is not "ready" for full rigor. (Even
the great rigorist Dedekind, in his *Continuity and Irrational Numbers*, says
that one must allow appeal to spatiotemporal intuition in introductory
calculus if one is not too lose too much time.)

Above this level, there may be more and more telescoping of deductive
steps, omission of citations of material supposed to be familiar, and other
shortcuts. Even in textbooks for undergraduate mathematics majors a number of steps may be—for legitimate pedagogical reasons, and not merely
on account of laziness on the part of the author—"left to the reader," but in
publications in specialist journals the tendency to leave things to the reader
is carried very much further. Generally, *definitions* are still expected to be
given in full, but *deductions* may be given only in outline, or anyhow, only
in what would *seem to a student rather than an expert* to be a mere outline.

The "Theoretical Mathematics" Controversy

Appreciation of this point is needed if the philosopher or student of philosophy is to get anything out of the controversy over rigor and proof that
appeared in the pages of the *BAMS* a few years back, and drew much attention in the mathematical community at the time (explicit public, published
discussion of such matters being very rare). The controversy began with
an article Jaffe and Quinn 1993 oddly titled "Theoretical Mathematics."
What the authors of the article, the mathematical physicist Arthur Jaffe,
and the geometric topologist Quinn, whom we have already met, meant
but were perhaps to polite to say was *speculative* or *heuristic* as opposed to
rigorous mathematics.

Though Jaffe and Quinn write with almost excessive academic polite-
ness, and though much of their polemic occurs on the level of subtext
rather than text, they managed to tread on quite a few toes, and the list
of eminent mathematicians who felt impelled to reply was a long one. It
will perhaps be best to begin, not with the original article that provoked
all these responses, but with the short, sober, sensible contribution of the
applied mathematician James Glimm (in Atiyah 1994, 6–7).

Glimm's main claims, to put them in my own words, are three. First,
applications cannot always wait for rigor. Second, this being so, the way to

honor the ideal of rigor in applied mathematics is by "honesty in advertising," calling theorems *theorems* and conjectures *conjectures*, proof *proof* and heuristics *heuristics*. Third, the community of applied mathematicians is actually quite good about doing this. The first two points seem to me indisputable, and I myself am not knowledgeable enough to dispute the third even if I were inclined to do so, which I am not. In saying all this, however, Glimm is not so much objecting, or even responding, to anything Jaffe and Quinn say, as making explicit the implicit presuppositions of their discussion.

Their positive point is precisely that more room ought to be made for speculative or heuristic mathematics, provided only that it is properly labeled as such. This is something hardly deniable given the very important contributions of Witten, the sole physicist ever to be awarded a Fields medal, who has put forward a host of important mathematical formulations, some of which have been, with varying degrees of difficulty, and with by-products of varying degrees of importance, rigorously proved by others, and others of which remain the object of intense investigation. And Jaffe and Quinn, in making their positive point, are not merely belatedly and grudgingly acknowledging what the achievements of Witten have made undeniable. On the contrary, Jaffe at least is closely identified with the editorial board of journal *Communications in Mathematical Physics*, which has been ahead of the pack in acknowledging that speculative or heuristic work has a legitimate and important role in mathematics.

There is, however, a negative point accompanying Jaffe's and Quinn's positive point, and something of a quarrel (not with any applied mathematicians but) with certain pure mathematicians who, they seem to insinuate, are *not* behaving the way Glimm says applied mathematicians do, but rather are putting forward speculative or heuristic mathematics *as if it were rigorous mathematics*. They are claiming the honor of and credit for proving theorems while actually producing at best sketches of proofs or strategies for a proof, leaving it to someone else to clean up after them and produce an honest proof. (I am putting things a bit more bluntly than Jaffe and Quinn themselves do.)

They propose a new labeling system—the specifics of which may be and are disputed even among those who approve the general idea—in which speculative results would be marked as speculative and rigorous as rigorous, and above all they advocate reserving a substantial share of honor and credit for those who do the hard work required to move a result from the former to the latter column. The reader who is familiar with the work and reputation of the various writers who join Glimm in responding will find much of what they have to say amusing as well as enlightening. A few produce under the provocation of the Jaffe–Quinn article exaggerated versions of things they have said more soberly elsewhere, amounting in the case of the hilarious letter from the catastrophe theorist René Thom (Atiyah 1994, 25–6) to a virtual parody of himself.

The most substantial, serious, sustained reply comes from a well-known Fields medalist, the unfortunately recently prematurely deceased William Thurston (1994), who seems to have been, though Jaffe and Quinn mention him only in one brief paragraph, one of the more important targets of their negative observations. In their original article, they speak of what is perhaps Thurston's most famous contribution as a "grand insight delivered with beautiful but insufficient hints, the proof [of which] was never fully published" (Jaffe and Quinn 1993, 8).

One thing Thurston brings out is the point I was making earlier, about what I identified as the element of truth in the slogan "a proof is what convinces," namely, that what may not suffice as a proof for one audience may count as such to another, more expert audience. You have to know a good deal about an advanced mathematical subject to be able to follow proofs written for experts in that subject, and perhaps not all his critics know enough. (I am putting things a little more bluntly than Thurston does.)

I do not wish to say much about this point. On the one hand, it is entirely safe to say that doubtless some mathematicians do put forward conjectures and heuristics as if they were theorems and proofs, while doubtless other mathematicians fail to follow perfectly good proofs of perfectly good theorems, mistaking rigor for speculation, simply because they lack sufficient background expertise. On the other hand, while it is safe to say all this while speaking generally, it would be both impudent and imprudent for a mere philosopher or logician to try to name names, and point to any specific eminent mathematician as an offender of the one kind or the other.

Rather than prolong what would inevitably be an inadequate attempt to restate Thurston's sometimes subtle points in my own words, let me simply commend the whole exchange of which his contribution is an important part to the reader's attention. The more background knowledge of the personalities and subfields involved one brings to the reading of the "theoretical mathematics" debate, the more one will get out of it, but even the reader with minimal technical prerequisites will find a great deal of illumination in it. Most of the writers, indeed, go out of their way to avoid technicalities, or at least to assure the reader that the points they are making do not depend on any deep familiarity with them.

One more point, at least indirectly relevant to the debate, is perhaps worth underscoring, namely, that there is a certain element of *falsehood* in "a proof is what convinces" and related formulations. The point is not the one I made earlier, that sometimes something convinces without amounting to a proof, but the reverse one that sometimes something amounts to a proof without convincing. And the point is not *just* the one we saw Hume making, that a proof may fail to convince because of our general awareness of human fallibility. The point is that worries about human fallibility are more pressing with some specific kinds of proofs than others.

A purported proof is much more likely to contain difficult-to-detect irreparable errors, and so not really amount to a genuine proof, if it is presented simply as a sequence of deductive steps, without enough hints as to overall strategy. This is *why*, for an expert, what looks to a non-expert like a mere outline of a proof may be more compelling than a lengthier account with a lot of fussy details, if the fussy details obscure the overall strategy.

Now no participant in the "theoretical mathematics" debate, as I understand it, is saying that a purported proof that includes no hint of the overall strategy of argument, nor clues as to how it was discovered, and contains perhaps instead an excessive amount of fussy deductive detail better left to the reader—a proof that provides no map of the woods, but only shows us the trees, including even the smallest and least important saplings—is a *non*proof. It *may* be a rigorous proof, containing no gaps or fallacies, but without a map of the woods we will have difficulty convincing ourselves that it is.

Moreover, *even if we are eventually convinced, the proof will still be unsatisfying*. For one has to acknowledge that there are dimensions of assessment of proofs that have little or nothing to do with rigor. There can be proofs that are genuine, rigorous proofs but lack the feature found in the "best" proofs in mathematics, of introducing methods applicable not merely to the specific problem at hand, but to a wide range of further problems, and of framing along the way definitions that isolate notions of importance not only in the original context but in many others.

There can be proofs that are genuine, rigorous proofs but nonetheless are "unenlightening" or "unexplanatory." Not every logically cogent, deductively rigorous proof is a *good* proof in the fullest sense.

I had an English teacher in high school who simply would not admit the existence of bad poetry. For him, not even Joyce Kilmer's "Trees" was a bad poem; rather, it was mere "verse" and not genuine "poetry." One must not use the word "proof" the way my teacher used "poem." There can be unsatisfying, unenlightening, unexplanatory proofs as well as satisfying, enlightening, explanatory ones. The distinction is important, but an analysis of mathematical "enlighteningness" or "explanatoriness" is no part of my present project of analyzing "proof."[22]

The most serious problem with the slogan "a proof is what convinces" remains the one we first noted, that what convinces need not amount to a proof. This problem is partly met by the Bass variant: "A proof is what convinces the mathematician that a formal proof exists," taking a formal proof here to mean something like a deduction in which every inferential step is a fairly simple one recognized as valid by essentially all logic texts, without necessarily requiring "formalization" in the sense of transcription into artificial symbolic notation. Testimonies by expert witnesses to the effect that a result is true, nonverbal thought, analogical extrapolation, generic reasoning, sudden insight, inductive generalization, spatiotemporal intuition, may all help convince us that a result is *true*, but they provide little direct support for a belief that it is *formally provable*.

They especially fail to supply such support to a reader who is aware of Gödel's celebrated *incompleteness theorem*, according to which, whatever postulates or axioms one begins from (so long as they are internally self-consistent and not self-contradictory), there will be mathematical truths that cannot be formally proved from them.

The Bass variant also gives some insight into what the debates over "theoretical mathematics" are about: No one is saying that formal proofs should actually be given; the debates are over what it takes to show that it

---

[22] Many self-identified "philosophers of mathematical practice" have attempted to analyze what "enlightening" or "explanatory" can mean in application to proofs, without any consensus emerging among them so far. The task they have assigned themselves is no easy one. For such honorific adjectives are, unlike technical mathematical terms, sufficiently imprecise in their meaning that many mathematicians may use them to mean little more than "of the kind I like." And different kinds of mathematicians like different kinds of proofs.

would in principle be possible to do so. But even the Bass variant does not fully meet the problem of characterizing proof and rigor, even if one sets aside worries about proofs that do not convince, and is only concerned with conviction being produced by means other than by proof. For after all, testimonies by expert witnesses to the effect (not just that a result is *true*, but) that it is *formally provable*, do tend to convince one that a formal proof exists, though presumably mere testimonies never amount, however numerous and eminent the figures supplying them, to something that can properly be called "proof."

To be a genuine proof, a supporting argument must not only produce conviction that a formal proof exists, but it must do so *in the right way*. But what *is* the right way? Can we say anything more than that it is not the way of mere testimony? I must confess myself to be unable to give a full answer—which is hardly surprising if, as the debate over "theoretical mathematics" suggests, mathematicians at the highest level themselves are not wholly agreed. Let me tentatively suggest, as a proposal for further examination rather than immediate adoption, that "the right way" to produce conviction that a formal proof exists is primarily by *supplying enough steps* of the purported formal proof.[23]

Here the Kac variant, "A rigorous proof is what convinces even an unreasonable person," comes to mind. Perhaps in principle nothing less than a complete breakdown into *obvious* steps will do, and what will do in practice is enough steps to produce conviction that *if pressed further by an unreasonably persistent questioner*, things could be broken down to this point.[24] That, at any rate, is my final suggestion: What rigor requires is that each new result should be obtained from earlier results by presenting enough deductive steps to produce conviction that a full breakdown into obvious deductive steps would in principle be possible. Lots of room for disagreement, even among experts, is contained in that word "enough," as is room for recognition that what is enough for one audience may not

---

[23] In this connection, the discussion in Steiner 1975 (ch. 3, sect. 3) of the philosopher or logician as a "midwife" who, by persistently demanding more and more intermediate steps, finally extracts something like a fully detailed, formal proof from the mathematician, is highly illuminating.

[24] Even with this formulation, there may be room for quibbling about "obvious." For recall the old anecdote about Norbert Weiner. According to this probably apocryphal tale, in a classroom lecture he once said that some step was obvious, and when questioned by a student, stood and pondered a long while, went off to his office and scribbled for several minutes, and finally returned to the classroom to say, "Yes, it is obvious."

be enough for another.[25] Even acknowledging that the proposed formulation would doubtless require any number of refinements, I think it will be
enough to provide some guidance in certain cases where there may be a
degree of uncertainty. At any rate, if it is deficient, its deficiencies will be
most likely to become apparent if we try to apply it in hard cases.

Theorems and Theorems* and Theorems†

The kinds of cases I have in mind here are *not* those that may be in dispute between Jaffe and Quinn on the one side and Thurston on the other.
Those cases, as I have said, the prudent philosopher or logician must leave
to the mathematical community. The kinds of cases I have in mind are,
rather, those involving such phenomena as so-called *computer-dependent
proofs* and *zero-knowledge proofs*. The terminology, obviously, is
question-begging, obliging us to pose the issues under dispute in seemingly tautological forms such as "Is a computer-dependent proof a proof?"
Let us not allow ourselves to be distracted by this feature of the dialectical situation, since the labels involved, however tendentious, are well
established.

However much Jaffe and Quinn, Thurston, and other participants in the
BAMS debate may differ among themselves about what constitutes speculative or non-rigorous mathematics, and what role such mathematics
should be allowed to play, I take it none of them, except perhaps Thom or
the late Benoît Mandelbrot, agrees with the sharply anti-rigorist claims of
Scientific American writer John Horgan in his notorious 1993 article "The

---

[25] One thing the proposed formulation does *not* seem to leave room for is a role for *diagrams*, such as one can find in almost any issue of almost any current journal in the library:
Their inclusion may play a part in convincing readers that a formal proof exists, even though
diagrams are not themselves deductive steps in a formal proofs. The role of diagrams was
an issue even in connection with Euclid's geometry, where the diagrams that accompany
each proof seldom contain information not conveyed in words in the text. Today the kinds
of diagrams in use range from computer graphics picturing geometric entities to charts of
logical relationships. Some are surely no more essential to the proofs in whose midst they
appear than the illustrations that accompanied many Victorian novels on first publication
were essential to the literary value of those works. Other diagrams may play a more important role, especially perhaps in those types of abstract algebra where what is called "diagram
chasing" is ubiquitous; though even here, the kinds of diagrams that are chased seem only to
abbreviate information that could be put, more cumbersomely and in a way less easy to take
in, in words. But the issue is a large one, deserving a monograph of its own, and I will have to
set it aside.

Death of Proof." According to this piece, written not long after Wiles's solution to the Fermat problem was announced, we were already then on the cusp of a Kuhnian paradigm shift. Wiles's proof of the Fermat conjecture was the last gasp of a dying mathematical culture. Soon the older generation would die off, and theorem-proving would be cast into the dustbin of history. A new cyber-savvy generation would take over, to create a brave new world free from neurotic hang-ups over rigor and cured of an obsession with justifying discoveries by deductive proof.

The Horgan piece was typical of the hype that surrounds everything connected with computers, even if the account I have just given of it is exaggerated and something of a parody. Certainly, computers have changed things. On the one hand, they have in many ways made the mathematician's life easier: One has to make fewer trips to the library, as so much of the literature is now available on line; collaboration with partners whom one seldom gets to see in person is also greatly facilitated, as is dissemination of results independently of journals. On the other hand, the burden of typesetting has been transferred from publishers to mathematicians, who are now practically required to submit their work in TEX format. But as for proofs being replaced by computer experiments, that still isn't a widespread trend. Proof may be on the way out, as certain futurologists predict, but if so it is departing at a slow and stately pace, and most of the major centers of mathematical research may be washed away by the melting polar ice caps before proof has entirely disappeared from the major mathematics journals.[26]

---

[26] The best-known advocate of computer-era anti-rigorism, Doron Zeilberger (see his archive 1995–2014), though he may be confident that what Horgan has predicted will eventually come to pass, seems dissatisfied with the pace at which the transition is occurring. The philosophical reader who browses among his opinion pieces will find many both entertaining and informative, but needs to be aware that his remains at present very much a minority point of view: Zeilberger is a gadfly, and delights in that role. (The reader must also carefully note whether the date of a given piece is April 1.) Zeilberger formerly advocated a kind of *semi*-rigorism, turning on algorithmic "proofs" of a kind discussed in his 1994 position paper "Theorems for a Price." What was at issue here was *not* the stunt pulled by the University of Edinburgh, which offered to sell theorems produced by one of its automated theorem-proving devices, in the sense of selling the right to *name* the theorems in question, but rather an extension of existing work in which computers prove mathematical identities of a certain important class, to work in which the computers would not produce proofs that a given identity holds, but rather would establish with high probability that the identity holds. He foresaw the kind of mathematics devoted to rigorous *human* theorem-proving dwindling in the not-too-distant future to the activity of a small group of harmless eccentrics, but what he foresaw as replacing it was a kind of mathematics in which an abstract of a paper might

And yet the Horgan piece and others like it, despite their exaggerations, do have the value of calling attention to certain phenomena in the world of mathematics that, however marginal they may remain even today, well into the twenty-first century, are of considerable interest to the philosopher— especially since for philosophers contemplation of how mathematics *potentially could be* done is as relevant as speculation on how mathematics *actually will be* done. Computer-dependent and zero-knowledge "proofs" are major cases in point, though there is a difference between the two, in that computer-dependent proofs are an actual phenomenon, while zero-knowledge proofs are more in the nature of a conceivable future development. Let me take the two phenomena up in the order listed.

The present situation in mainstream pure mathematics is that rigorous proof is demanded. Though there is a great deal of heuristic evidence in favor of the Riemann hypothesis, making the conjecture plausible enough that it seems worthwhile to begin exploring its consequences even while a proof is lacking, still the Riemann hypothesis is not called a "theorem," and neither are the results deduced from it. Such results have to be asserted in *conditional* or *hypothetical* form: "If the Riemann hypothesis holds, then . . ."[27] This much seems uncontentious. What is more controversial is whether computer-dependent proofs still count as having only some kind of "conditional" or "probationary" status.

Let us first get clear what is *not* at issue. What is at issue here is not the role of computer experimentation in suggesting *conjectures*, or by suggest- ing conjectured lemmas on the way to conjectured theorems suggesting *proof-strategies*. These matters clearly pertain to the context of discovery rather than of justification. Nor is what is at issue the use of "proof-assistants" (of which more shortly). When I speak of "computer-*dependent*" proofs, I have in mind situations where an assertion is supported by the report

announce that, in his words (1994, 6), "in a certain precise sense, . . . the Goldbach conjecture is true with probability larger than 0.99999, and . . . its complete truth could be determined with a budget of $10B." I will not be discussing his algorithmic "proofs" in more detail, since he has subsequently given up semi-rigorism for outright anti-rigorism.

[27]   An important early example of such a result occurs in the work of Skewes, mentioned earlier. He obtained a much better upper bound with the assumption of the Riemann hypoth- esis than without it. A situation parallel to that with the Riemann hypothesis in pure math- ematics—a situation where the experts are so sure of the truth of an unproved result that it seems worth taking the risk of working on exploring its consequences, pointless though such work would become were the conjecture ever, contrary to expectation, disproved—obtains with regard to the $P \neq NP$ conjecture in theoretical computer science.

that extensive computer calculations, *much too long to be printed out, or if printed out taken in and evaluated by a human reader,* show that the assertion holds. The premier example was the 1997 proof of the four-color theorem by Kenneth Appel and Wolfgang Haken.[28] The issue whether in such cases the "proof" is really a proof and the "theorem" really a theorem is one on which there has been genuine indecision and difference of opinion among mathematicians. The issue is not whether one has an "enlightening or "explanatory" proof, since it is unanimously agreed that one doesn't; the issue is whether what one has should be accounted a proof at all.

At one end of the spectrum of opinion, when the work of Appel-Haken first emerged, there were those who simply rejected this kind of reliance on computers, admitting as "proofs" only humanly checkable proofs. At the other end of the spectrum, there are those who saw no really novel issue here, finding little difference between this situation and one where a mathematician at some point in a proof uses a calculator rather than doing some complicated arithmetic by hand. For while doing computations, human mathematicians in effect turn themselves into machines insofar as while calculating one is not thinking about the *meaning* of the symbols being manipulated, but only about their *shape*.

An intermediate position would demand that the *program* followed by the machine must be exhibited, and some kind of assurance given that, if this software is accurately implemented by the hardware, then the result obtained will be correct. If this is done we can check the accuracy of any one piece of hardware by running the same software on another piece of hardware, just as in the case of humanly feasible calculations we can check the accuracy of one human calculator by having the same computations done by another human calculator. Without a verification that the software is in order, however, checking by having the same computations done on another machine would be (to borrow an image from Wittgenstein) like checking the correctness of a news report by buying another copy of the same newspaper. Opinion may be converging on something like this intermediate view, among those who think about the issue at all, but it cannot be said that at present there exists any agreed protocol concerning the documentation of programs, or program-verification.

[28] See Rademacher and Toeplitz 1966, ch. 12, for an account of the four-color question as it stood before Appel and Haken's work, including the easier five-color theorem, and Appel and Haken 1989 for the full, final version of their work, including corrections to some flaws in the original version.

The situation has been changed by the emergence of the computerized "proof-assistants," alluded to a moment ago, from earlier work on "automated theorem-proving." The older work involved computers *on their own* searching for fully formalized proofs, written in a formal language; the newer work involves the production of such proofs by *interaction* with a human mathematician, which can take one much further much faster. Much pioneering work on such assistants took place among computer scientists, rather than core mathematicians, and concerned precisely the difficult task just mentioned of program-verification. It has since been applied to more centrally mathematical questions, including the four-color theorem itself.[29] A proof-assistant not only helps with the construction of a formal proof, but also verifies that it is formally correct at the same time. That the program for the proof-assistant behaves as advertised is itself something that ought to be verified, but that is not a question specific to any particular result that the proof-assistant is used to prove.

The development of suitable assistants is at present a growth industry. There exist several rival formats (Mizar and Isabelle and Coq, to name three), each capable of different implementations. These differ in the degree to which they are fully automated, and in the ease with which the formal proofs they produce are readable by human mathematicians, with perhaps an inverse relationship between the two factors. No single approach has yet achieved universal acceptance, and perhaps none ever will (as with operating systems). But already the existence of such assistants has led to the creation of substantial archives of formal proofs, even in areas where the production of such proofs, though presumed theoretically possible, was previously supposed practically infeasible.

This represents an influence of computers on the practice of mathematics diametrically opposite to what was earlier suggested by chatter about "the death of proof." Whereas anti-rigorists have for some time been predicting that sooner or later all mathematical journals will eventually be given over to experimental mathematics, abandoning requirements of proof, one now hears opposing predictions that sooner or later all mathematical journals will require formal proofs, or computer code

---

[29] In the 1990s, there were improvements on Appel and Haken by others, reducing the complexity of the proof, though not quite to the point of producing a complete proof that it would be feasible for a human being to check by hand. Gonthier 2008 announces a complete formal proof. We have probably not heard the last word on this question.

for generating them. A bifurcation or trifurcation of mathematical practice is also conceivable, rather than the triumph of either of the two computer-influenced trends, or the persistence of the present situation, where proofs are demanded, but not formal proofs.

Time will tell, and I make no prediction myself. Many of us who belong to the generation old enough to be able to remember how circa 1960 we were being promised that real-time mechanical translation was just around the corner may regard any predictions that there will be a major change in mathematical practice in the short-term future with some degree of skepticism. But, of course, those of us old enough to remember 1960 cannot expect to be on the scene for very far into the longer-term future.

One main policy question would be whether computer-dependent proofs have to be identified as such, or rather—since publication of a proof is demanded before one gets credit for a theorem, and if the proof is computer-dependent, that fact will be evident from the published version—whether *later* results that *depend on* earlier results established by computer-dependent proofs have to be identified as such, have to be marked off as indirectly computer-dependent. Only if this is done will computer-independent mathematics be clearly marked off from computer-dependent mathematics. And as computer-dependent proofs multiply, only if such marking off is insisted upon will computer-independent mathematics continue to exist as a readily identifiable activity.

So far, perhaps because the phenomenon of computer-dependent proof, despite its having been around for some while now, and despite its becoming rather more common from year to year, is still comparatively new and rare, no clear policy has emerged about the extent to which a piece of mathematics that itself contains no computer-dependent steps, but that does cite earlier results whose proofs were computer-dependent, needs to be specially marked. I have looked at the instructions for authors on the inside covers of several prestigious journals, and found no special instructions on this point. Certainly, no uniform theorem-starring-convention has emerged. Perhaps one never will, for the mathematical community seems to be tending in the direction of simply accepting theorems with computer-dependent proofs on a par with other theorems, provided the programs are adequately documented.

The case seems different with zero-knowledge proofs, or *zero-knowledge protocols* as they are alternatively and less tendentiously called. Not to go into technical details, such a protocol is a procedure by which party X can convince party Y, to an arbitrarily high degree of probability, that party X is in possession of a proof of a certain result, without revealing to party Y anything about the content of the proof in question. Zero-knowledge protocols have actual practical applications in connection with authentication problems, but the mathematical results involved in such applications are too specific to deserve to be called "theorems." (One doesn't call a specific numerical equation such as $3^2 + 4^2 = 5^2$ a "theorem"; nor, to use an example of Hardy's, does one call the solution to a specific chess problem a "theorem," though such problems are essentially mathematical.) The limited range thus far of the zero-knowledge phenomenon is in part owing to the fact that zero-knowledge protocols can only be applied where the concealed proof is a detailed, formal one.

It is nonetheless not too difficult to conceive of circumstances in which "theorems" worthy of the title would come into circulation with only zero-knowledge proofs. In the national-security state, scientific results are sometimes "classified," and their publication banned. Certain kinds of mathematical theorems, relevant to cryptography, might well be (and according to rumor have been) subjected to such bans. One can conceive of circumstances in which the government officials involved might judge that the announcement of a certain theorem would by itself be harmless or even useful, but that the public must not be allowed to know of certain techniques employed in its proof.

It hardly needs saying that the mere announcement by a government official that such-and-such a theorem holds would be unlikely to result in general acceptance in the mathematical community. After all, government officials have been known to engage in campaigns of deliberate "disinformation." A zero-knowledge proof could in such a case perhaps help win acceptance for the result, by convincing mathematicians that a proof does exist, even if they are not going to be shown it. There are, after all, some important results that have come to be widely accepted though publication of their proofs has been very long delayed.

A different kind of change in mathematical culture, not involving government interference, can also be imagined, if one is engaging in science fiction, as leading to a proliferation of zero-knowledge proofs of substantial theorems. In the days of Cardano, mathematicians seem to have viewed each other less as colleagues than as rivals, and in some cases less as rivals than as mortal enemies. Advances in technique were not published, but rather were concealed so as to maintain a competitive advantage in public problem-solving contests. Society may, perhaps, be tending in the unpleasant direction of a return to this sort of situation, in which case zero-knowledge proofs might be just what was wanted.

The question is, if this sort of thing became common, would mathematicians insist that new results depending on old results for which only zero-knowledge proofs had been given should be marked with some special symbol? No one can predict such things with confidence, but I would guess that many mathematicians would want to see cloak-and-dagger results marked off somehow, with a star or perhaps better with a dagger (†). My inclination is to say that, while computer-dependent proofs may be proofs, zero-knowledge proofs are not: that giving the kind of performance involved in a zero-knowledge protocol does not amount to *giving* a proof, however convincingly it establishes that one *has* a proof.

A common feature of computer-dependent and zero-knowledge "proofs" is that a very high degree of rational belief is produced not merely that the result is true, but that a formal proof exists. So the Bass criterion in its original form is met, though not the proposed amendment, that the conviction must be produced by exhibiting enough steps of such a formal proof. Yet I find myself inclined to say that a computer-dependent proof *should* (provided the program is suitably documented), while a zero-knowledge should *not* count as a "proof." If so, then we have still not got a clear necessary and sufficient condition for proofhood. I will, nonetheless, at this point set aside the topic of rigor, encouraging the interested reader to press on with it.[30] As for what the mathematical community of the future will decide, as computer-dependent proofs and/or zero-knowledge and/or other phenomena at this time still over the horizon come to play a substantial role as well, recall Weyl's null hypothesis, that the decisions of the mathematical community may "defy complete objective rationalization."

---

[30] Let me especially recommend Krantz 2011. Besides going over some of the same ground as the first two chapters of the present work, but from the point of view of a professional mathematician rather than a logician and philosopher, Krantz offers further case studies worth pondering, accessible with only a little more mathematical background than I have wanted to assume here. Noteworthy is the treatment of three cases considered in his ch. 10: Louis de Brange and the Bieberbach Conjecture; Wu-Yi Hsiang and the Kepler Problem; and Grisha Perelman and the Poincaré Conjecture. Krantz was, by the way, with his 1994 reply to Horgan, very much the public face of the mathematics profession during the controversy over the alleged "death of proof." This circumstance makes his own predictions at the end of his book as to what will be accepted as a proof a century hence of especial interest.

3
Structure and Structuralism

Bourbaki and Structures

If a purported proof is a genuine proof—if it does not contain gaps or flaws that, if noticed by the mathematical community, would be considered to vitiate the alleged proof and demote the "theorem" claimed to have been proved back to the status of "conjecture"—then it should be possible in principle to elaborate it, by filling in steps and other editorial changes, into a formal proof from the generally accepted axioms of set theory, Zermelo-Fraenkel set theory with choice (ZFC). So much has been suggested already. And it has been added that no one expects working mathematicians to produce actual formal proofs. And it has been admitted that the question of how close the mathematician has to come to doing so before a result can be claimed to have been proved has been left without complete resolution, as an issue in part still under debate among professional mathematicians, and one not unlikely to be affected by continuing technological developments.

Since so many mathematicians have at best foggy ideas as to what, apart from the notorious axiom of choice, the axioms of set theory amount to, it is doubtful how many would commit themselves to the foregoing statement that a genuine proof can be elaborated into a deduction from those axioms. A statement that I expect would be more likely to gain widespread assent would be, rather, that if a purported proof is a genuine proof, then it should be possible in principle to elaborate it into a proof in the style of Bourbaki's *Éléments de mathématique*. It is time to say something about that great but forever unfinished series.

The group of post-Poincaré French rigorists writing under the collective pseudonym "Bourbaki" chose the title of their series of publications with care: "*Mathématique*," singular, in place of the normal French "*mathématiques*," plural, reflects a strong commitment to the unity of mathematics;

"*éléments*" reflects the ambition to produce a modern counterpart of Euclid's *Elements*, a paradigm of modern rigor. That it should be possible in principle for a Bourbaki-style proof to be elaborated into a formal proof from the axioms of set theory is what the Bourbachistes (as members of the group are called) themselves wished to claim, and set theorists and logicians seem to agree. Bourbaki was certainly not the originator of the idea that the whole of modern mathematics can be developed on the basis of axiomatic set theory, but no one has produced a more important and influential attempt at a systematic implementation of the idea, a systematic codification of modern mathematics on a set-theoretic foundation, or in a set-theoretic framework. This is in a way remarkable, since Bourbaki evinces less than no interest in set theory as a subject in its own right.

The first published volume (1939) is a short pamphlet that opens by telling the reader that it contains all the definitions and results from set theory that will be used in subsequent volumes, but no proofs. A full volume with proofs (1957) appeared only after more than a dozen other volumes had come out. The initial omission of proofs made it unnecessary to give a list of ultimate axioms. Public statements by individual Bourbachistes, and by the whole group in collective publications under the name of Bourbaki,[1] claimed that what are essentially Zermelo's axioms suffice. What of Fraenkel's axiom of replacement? It is indicative of how far Bourbaki was from engagement with the work of set theorists that it is not mentioned. It would need to be added to their framework if this is to accommodate the whole of orthodox mathematics, and indeed it is added in later. With this amendment, the statement with which we began may be restated as a double claim: that it is possible to elaborate any genuine proof into a bourbachique proof, and any bourbachique proof into a formal proof in ZFC.

The culmination of the volume is a discussion of "structures," and it is here, rather than in set-theoretic "foundations," that Bourbaki's real interests lie, though in practice in later volumes not much actual later use is made of the general notions introduced. The word "structure" is a

---

[1] Relevant public statements include "Foundations of Mathematics for the Working Mathematician," written in English for the *Journal of Symbolic Logic* (1949), and "The Architecture of Mathematics," published in English translation in the *American Mathematical Monthly* (1950). Bourbaki has been severely taken to task for his confusions over Zermelo *versus* Zermelo-Fraenkel, and more generally for his ignoring developments in logic from Gödel onwards, by the set theorist Adrian Mathias, in an 1992 article bluntly titled "The Ignorance of Bourbaki."

common label for what I have heretofore rather awkwardly been calling "spaces or number systems or whatever," a label for the genus of which *groups, ordered sets, topological spaces,* and more are species. Bourbaki more than anyone popularized the use of this term, though its use in the specific technical sense Bourbaki eventually gives the word did not become common.

A structure has a set of elements, though structures are not just sets, but rather sets "with additional structure." In the case of groups, the "additional structure" is a distinguished two-place operation, called the *multiplication* operation of the group (or if the group is Abelian, which is to say, if the operation obeys the commutative law, the *addition* operation of the group). In the case of an ordered set, it is a distinguished two-place relation, called the *less-than* relation of the ordered set. In the case of a topological space, it is a family of distinguished subsets, called the *open* subsets of the topological space. A "structure" in Bourbaki's sense is a set together with such additional apparatus, though the most common admitted "abuse of language" is to use the same name for the structure as for the underlying set. Actually, there may be *two* or more sets, or domains of elements, as with a vector space, where there are both vectors and scalars, and sometimes there is yet other apparatus. Bourbaki's official general definition was complicated—far too much so to be reproduced here—because it was supposed to cover all cases.

Not every structure consisting of a set and a multiplication operation is a group, but only those for which certain "axioms of group theory" hold; and similarly for ordered sets and topological spaces. Here we have the situation Hodges was referring to when we quoted him as saying that "axioms" in mathematics today (except those of set theory!) are simply definitions of kinds of set-theoretic structures.

Closely connected with the notion of "structure" is the notion of a "structure-preserving map" between different structures. Actually, this phrase may be understood in either of two senses, a weaker one, leading to the notion of *homomorphism,* and a stronger one, leading to the notion of *isomorphism.* Only the latter was treated by Bourbaki in the volume in question, and only it will be of immediate concern here. An isomorphism is first of all a *correspondence* between the underlying sets of two structures. Each element of the one set, call it *A*, is associated with a unique element of the other set, call it *B*, and vice versa. Then there is the further condition that the association or mapping should "respect" the additional structure. What this condition amounts to depends on whether the "extra structure" is an operation or a relation or a family of subsets or whatever. It is one of Bourbaki's aims to formulate a general definition, but since details would take us too far afield, let me simply illustrate by treating the case where it is a two-place relation.

So let us be given a first structure consisting of a set A together with a relation R_A and a second structure consisting of a set B together with a relation R_B. What is further required of correspondence between A and B in order for it to count as an isomorphism? Just this, that if a_1 and a_2 correspond to b_1 and b_2, respectively, then we must have $a_1\,R_A\,a_2$ if and only if $b_1\,R_B\,b_2$. For example, if A and B and C are the sets

$$A = \{0,1,2,3,\ldots\}$$
$$B = \{1,2,4,8,\ldots\}$$
$$C = \{\ldots,1/8,1/4,1/2,1\}$$

of natural numbers, of powers of two, and of inverses of powers of two, and if the relation in each case is the usual order relation, then (i) the mapping associating an element a of A with the element 2^a of B is an isomorphism between the first and the second ordered set because bigger exponents give bigger powers; but (ii) the mapping associating it with the element $1/2^a$ of C is not an isomorphism, because order relations are reversed. Two structures are *isomorphic* if there is an isomorphism between them: Thus the A- and B-orders are isomorphic.

Isomorphic structures may be called "like-structured" (that is, nearly enough, what *isomorphic* means etymologically), and may be said to exhibit a common "structure" in a second sense of structure: what "structures" in the first sense have in common when there is a "structure-preserving" map between them. To prevent ambiguity, the second sense of "structure" can be avoided in favor of the more technical term "isomorphism type."

By contrast to the situation with the A- and B-orders, the A- and C-orders are *not* isomorphic. This does not, of course, follow just from the fact that the one correspondence between them that we considered turned out not to be an isomorphism, for this in itself does not exclude the possibility that there is not some other mapping that *is* an isomorphism. But there is indeed no such isomorphism. This is because, looking at the definition of isomorphism, we see that any isomorphism will have to associate the least element in one order (if it has one) with a least element in the other order (which must therefore have one, too), and the greatest element in one order (if it has one) with a greatest element in the other order (which must therefore have one, too). But the A-order has a least element (0) and no greatest, while the C-order has a greatest (1) and no least.

This last argument illustrates the fact that any statement that involves only the distinguished relation and the logical notions of negation ("not"), conjunction ("and"), disjunction ("or"), conditional ("if"), and universal and existential quantification ("all" and "some"), whether "first-order" (over elements x of the structure) or "second-order" (over subsets X of the structure), and that is true in any one structure, will automatically be true in any isomorphic structure. Pre-theoretically this is plausible, and with Tarski's technical definition of truth for languages involving such limited forms of expression it is provable, and for all kinds of structures, those with distinguished operations as much as those with distinguished relations.

The statement that there exists a least element is of this kind; indeed, *all* the following five statements are of this kind:

Trichotomy: *For any x and y exactly one of the following holds: x is less than y, or x is equal to y, or y is less than x.*

Transitivity: *For any x and y and z, if x is less than y and y is less than z, then x is less than z.*

Zero: *There is a* least *element, an x such that for every y other than x itself, x is less than y.*

Successor: *For every x there is a* next *element, a y such that x is less than y but there exists no z such that x is less than z and z is less than y.*

Induction: *For any set X of elements, if the least element belongs to X and the next element after any element belonging to X belongs to X, then all elements belong to X.*

All five happen to be true for the isomorphic A- and B-orders, but only the first two in the C-order. Dedekind in effect showed that *any* structures for which all five hold are isomorphic to each other.

In general, every list of statements of this kind, taken as "axioms," and together called a "theory," defines a class of structures, and one that will contain any structure isomorphic to any structure it contains. For instance, the class of ordered sets is determined by the theory consisting of the first two of the five axioms just listed. In some cases, like that of the theory consisting of *all* of the five axioms, the theory will determine a class in which *all* structures are isomorphic. Bourbaki calls the special case that of *univalent* theories, though the more usual word is *categorical*. Bourbaki calls the general case, where not all structures are isomorphic, *multivalent*. Bourbaki's first volume ends with the following remark:

The theory of integers, that of real numbers, classical Euclidean geometry, are univalent theories; the theory of ordered sets, the theory of groups, topology, are multivalent theories. The study of multivalent theories is the most striking feature distinguishing modern mathematics from classical mathematics.[2]

Even when the mathematician's primary interest is in the natural numbers, or real numbers, or Euclidean three-space, the characteristically modern way of studying such primary structures is to introduce secondary structures as auxiliaries and apply general theorems from the theory of ordered sets or groups or topological spaces or some other non-categorical or "multivalent" class of structures.

Bourbaki's anti-traditional conception of modern mathematics as concerned with "structures" has sometimes been called "structuralism"; but however novel it may have been when first enunciated, this kind of "structuralism" is now widely regarded as a trivial truism, so long as it is separated from the particulars of Bourbaki's attempt to pin down the relevant notion of "structure" in a technical definition. The contentious issue debated under the heading "structuralism" in contemporary philosophy of mathematics is not to be confused with Bourbaki's claim. (Still less is any form of structuralism to be discussed here to be confused with "structuralism" in the sense in which it is used in French intellectual life; connections between Bourbaki and Lévi-Strauss, for instance, are at best tenuous.)

The issue in contemporary philosophy of mathematics concerns a matter on which the quoted formulation from Bourbaki is a little unclear. Is a subject like number theory or analysis still to be thought of in a traditional way, as the study of a single structure, the natural number system or real number system as the case may be, the "intended interpretation" of the relevant theory? Or is it to be thought of now as the study of a class of isomorphic structures, all "models" of the theory, all structures for which

---

[2] *La théorie des nombres entiers, celle des nombres réels, la géométrie euclidienne classique, son des théories univalents; la théorie des ensembles ordonnés, la théorie des groupes, la topologie, la théorie des variétés différentielles, sont des théories multivalents. L'étude des théories multivalentes est le trait le plus frappant qui distingue la mathématique moderne de la mathématique classique.* Bourbaki 1939, 37.

the axioms are true? In the case of number theory, the relevant models (essentially, those of the five axioms) are called *progressions*, while in the case of analysis (where I will not display the rather longer list of axioms) the relevant models are called *complete ordered fields*.

What I have called the "paradox of rigor" can be restated in these terms as the remark that, if we make the study of the natural numbers or real numbers, say, completely rigorous, by reducing it to the deduction of consequences from appropriate axioms, all results will automatically become applicable to a whole class of structures, the class of models of those axioms. The underlying reason is the fact, already stated, that the same statements of the restricted kind under consideration are true in all models. Or rather, the reason is that fact plus the fact that seemingly all questions of mathematical interest, from Goldbach's conjecture in number theory to the Riemann hypothesis in analysis, are formulable by statements of the kind we have been discussing. Does there, then, remain any reason to be concerned with the natural numbers as distinguished from other progressions, or the real numbers distinguished from other complete ordered fields?

Are there such things as *the* natural number system or *the* real number system? This is what is at issue in contemporary philosophical debates. And what is the structuralist answer to the question? As it happens, there turn out to be two forms of structuralism, which answer the question differently. One says "no," the other says "yes, but … " I am, however, getting ahead of my story. Before I try to say anything more specific, a very rapid review of previous philosophy of mathematics may be desirable for readers with a background more mathematical than philosophical.

Philosophy and Mathematics

Philosophy originally encompassed almost all areas of intellectual inquiry, insofar as such inquiry appealed only to reasoned argument and the evidence of experience, and not revelation, tradition, faith, or authority. In the medieval university, the faculty of philosophy took in everything but the specialist professional studies of medicine, law, and theology; and so conservative are educational institutions that "doctor of philosophy" or Ph.D. remains to this day the name of the highest degree in molecular biology and comparative literature alike. The early modern period, roughly 1600–1800, saw the separation of what we now call "natural science" and was then called "natural philosophy" from the rest of philosophy. The term "physics," originally denoting a branch of philosophical speculation, became the name for a branch of scientific research. Virtually all the major philosophers of the period were either participants in this process, as in the case of the philosopher-mathematicians Descartes and Leibniz, or closely observed it in the hopes of being able to bring about a similar transformation of "moral philosophy" or philosophical "ethics."

The transition saw philosophy's vague commitment to "reasoned argument" converted into a demand for mathematical precision in enunciating hypotheses and deriving predictions from them (on the model of the most useful exemplars inherited from antiquity, in the work of Archimedes on floating bodies and on the lever), and the equally vague commitment to the "evidence of experience" into a demand for systematic observation (such as the astronomical tables from which Kepler started) and controlled experiment (such as Galileo, for one, conducted). There were inevitably some who put more emphasis on reason and mathematics, and others who put more emphasis on experience and experiment, and historians of philosophy from the nineteenth century on have conventionally divided early modern philosophers into two parties, the rationalists and the empiricists. The rationalists were at first tempted to imagine that physics could proceed in much the manner that mathematics seemed to proceed, simply by deduction from self-evident principles. The failure of Cartesian and the success of Newtonian physics shows that, while deduction from principles may still play a role, the principles must rest on observation and experiment, and not imagined "self-evidence." The empiricists were inclined to hold that *nothing* could be known without observation and experiment, and then were puzzled by the apparently exceptional case of mathematics.

We have already in passing encountered the terminology in which Kant, at the end of the early modern period, stated what the puzzle seemed to be. On the one hand, mathematics appeared to be a priori rather than a posteriori, meaning independent of rather than dependent on experience. On the other hand, mathematics appeared to be synthetic rather than analytic, meaning something that provides substantive knowledge of objects and not just formal knowledge of concepts. How synthetic a priori knowledge was possible in mathematics was for Kant the first question on the agenda of theoretical philosophy. Never before since Plato had philosophy of mathematics occupied quite so central a position in philosophy.

But we have already seen, in the discussion of non-Euclidean geometry, half the story of how Kant's presupposition that mathematics is synthetic a priori came to be challenged. During the period from Gauss to Einstein mathematical geometry came to be distinguished from physical geometry, and it came to be held that the latter is a posteriori (and the former is of the same character as arithmetic, algebra, and analysis, so that it presents no further philosophical problems not presented by those branches).

The other half of the story begins with Gottlob Frege, the founder of modern logic, writing at the height of the period of rigorization in the late nineteenth and early twentieth centuries. As Gauss had wished to reclassify geometry, not as synthetic a priori, but as a posteriori, so Frege wished to reclassify arithmetic (and algebra and analysis), not as synthetic a priori, but as analytic. For him this meant: deducible from suitable definitions by his expanded logic. Such is the thesis of *logicism*, which Russell continued to defend after discovering a fatal flaw in Frege's logic, introducing a revised logic of his own. The inconsistency of Frege's system, and the baroque complexity of Russell's, made them unsuitable as frameworks, but Russell's system was simplified by his student Ramsey. As mentioned in Chapter 2, Ramsey's simplified theory is just a little weaker than Zermelo's axiomatic set theory, which had already been introduced and was, with additions and amendments, ultimately to prevail.

Challenging the emerging consensus was Brouwer's intuitionism, already discussed. Kronecker's *finitism* had been a precursor, even more restrictive, and Martin-Löf's constructivism is a successor that survives as an active research program down to the present day. In the 1920s and 1930s, it became customary to speak of "three schools": logicism (and along with it the still-developing axiomatic set theory), intuitionism (not always at this period clearly distinguished from finitism), and formalism.

Of this last I have not yet said anything. It represented the response of Hilbert, the leading mathematician of the period, to the "foundational crisis." Hilbert's idea, not to put to fine a point on it, was to try to establish by means acceptable to finitists that axiomatic set theory is at least consistent. While the system to be proved consistent is ostensibly about infinite sets, the claim that the system is consistent amounts to a finite, combinatorial statement: From certain finite strings of symbols (the axioms of Zermelo's or whatever system) it is not possible by certain specified transformations (the rules of the system) to arrive at a certain other finite string (expressing that $0 = 1$). Whether this is so is a question perfectly meaningful even to adherents of very restrictive points to view, who do not believe that the strings in question (generally formulations of various claims mainly about "completed infinities") have any genuine meaning.

Hilbert's hope was to be able to convince them not only of the meaningfulness of the consistency statement, but of its truth. A kind of reliability can be shown finitistically to follow from the mere consistency of an infinitistic system: If an infinitistic system is consistent, its finitistically meaningful theorems will be true. Thus if consistency could be finitistically established, the legitimacy of the *instrumental* use of infinitistic mathematics could be established, and the infinitistic mathematician could give away everything to the finitist on the plane of philosophical principle, while giving away nothing on the plane of mathematical practice.[3]

The "three schools" period of interaction between philosophy and mathematics came to an end when Gödel proved that it was impossible to do what Hilbert was trying to do: to prove the consistency of a stronger system of axioms working in a weaker system of assumptions. Indeed, it is impossible to prove the consistency of any axiom system working within that system (unless the system is in fact *in*consistent, in which case it is possible to prove *anything* within it, even 0 = 1, or is so weak that no significant amount of mathematics can be done in it). More precisely, the period came to an end as, over the course of a decade or so, the implications of Gödel's work came to be assimilated.

The period of interaction that then ended had been notable for the involvement of leading figures in philosophy (Frege, Russell, Ramsey, Wittgenstein, Carnap) and mathematics (Poincaré, Brouwer, Hilbert, Weyl, Von Neumann) with each other's disciplines, to a degree that had not been seen since the days of Descartes and Leibniz, when some of the leading figures in philosophy actually were leading figures in mathematics as well, and to a degree that would be quite unthinkable today. It was notable also for the creation of a new field between mathematics and philosophy, that of modern logic, represented after Frege and Russell by Gödel, Tarski, Church, Turing, and many more.

---

[3] The thought that it might be useful for certain purposes to disregard the meanings of the strings and consider them only *as* strings subject to manipulation according to rules proved too subtle for some of Hilbert's critics, who accuse him of regarding mathematics as nothing but a meaningless game with symbols. Such an accusation betrays a gross ignorance or else a willful disregard of Hilbert's career and *œuvre*. The conventional label "formalist" goes better with this caricature than with Hilbert's actual views.

From the standpoint of logicians, the period was a glorious success, and ended with logic established as a flourishing field with four flourishing branches. *Set theory*, as a subject in its own right, concerned with transfinite cardinal and ordinal numbers and related matters, once put on an axiomatic basis by Zermelo, Fraenkel, and others, came to be wholly absorbed by logic. *Model theory* originated as a *general* theory of "structures," beginning with Tarski's rigorous definition of just what it means to say that a statement is true in a structure: It contrasts with the various branches of core mathematics that study particular *kinds* of structures or models, such as groups or topological spaces, though Abraham Robinson (whom we met earlier as the inventor of non-standard analysis) eventually gave the field an orientation towards applications to algebra. *Proof theory*, in the broadest sense, is concerned with the trade-offs between power and security that Gödel's incompleteness theorems show to be inevitable: Stronger axioms lead to more theorems, but at greater risk of contradiction. *Computability theory*, developed from germs in Gödel's work by Church, Turing, and others, provided the intellectual background to the development of the first electronic digital computers—needless to say, there were very substantial practical engineering problems involved as well—and from it sprang what is now the vast field of theoretical computer science.

From the point of view of mathematicians, the period succeeded at least in this, that from it emerged, if not a "foundation" for mathematics in the sense of a list of self-evident principles, at least a framework, that of axiomatic set theory, within which mathematicians could develop their subject rigorously, and whose consistency is a reasonable assumption. It is a reasonable assumption partly because there is arguably a plausible heuristic picture of a universe of sets associated with the formal axioms (one that will be briefly discussed at the end of this book), and partly because by now we have long experience of working with the axiom system in daring ways without falling into contradiction. And the fact that we have no *proof* of consistency is somehow less troubling once one knows from Gödel that there cannot be such a proof for *any* system strong enough to do any serious mathematics in—not a consistency proof based on weaker principles than those embodied in the system itself.

The aim of Hilbert had been "to get rid of once and for all" (*einfüralle-mal aus der Welt zu schaffen*) foundational questions in mathematics, and though Gödel showed this could not be done in the way Hilbert was trying to do it, the goal was nonetheless in some sense achieved. Mathematicians acquiesced in the set-theoretic framework, mostly without enthusiasm, doubtless in large part because the standard axiom system ZFC permits them to do pretty much everything that it would come naturally for them to do in the way of constructing new structures out of old, and permits the deduction of all the familiar principles of traditional mathematics that had been put on any sort of rigorous basis over the preceding century, and *does not require any fussy bookkeeping* to ensure that one is remaining within bounds going forward. In short, it permits the mathematician to *stop thinking about* foundational questions. This they generally have done, and to an extent that sometimes dismays logicians.

But is it not unacceptably authoritarian to impose a single framework? Should one not be more tolerant and accommodating, and "let a hundred flowers bloom"? This thought betrays a serious lack of appreciation of what so moved Bourbaki, the value of *formal* unification within an agreed framework as facilitating *material* unification of mathematics, by permitting the mathematician working in one branch to draw on results from the widest range of other branches, should they seem relevant, without having to worry that those results may ultimately be based on different and incompatible principles. And while holding to a unified framework, it is really not so difficult to make room in a way for a kind of pluralism.

Today, if someone wants to use assumptions that have not been proved within the accepted framework, such as the Riemann hypothesis (RH), or that are known to be unprovable within that framework, such as the continuum hypothesis (CH), there is no institutional obstacle at all to doing so. All that is really required is that the dependence of results should be clearly indicated in their statements. Instead of writing "Theorem: Let X be ... " and then using whatever hypothesis it may be in the proof, one writes "Theorem: Assume RH, and let X be ... " or "Theorem (CH): Let X be ...." A later worker wishing to make use of the result will see the annotation, and if not deterred by it, may proceed to deduce consequences from it. All that is required is *passing on the label* and marking the new results, too, by "Assume RH ... " or "(CH)," or whatever. Is this too much to ask?

But what if someone does not wish to *add* to the orthodox framework, but to *subtract* from it, say by imposing constructivist or predicativist limitations? Here accommodation is provided in a different way. Instead of just stating "Theorem: Let X be ... " and giving a proof by restricted means, one states a so-called metatheorem, thus: "The following is provable in

Martin-Löf's constructive type theory: Let X be . . ." or "The following is provable in Feferman's predicative analysis: Let X be . . . ." So long as the background constructivist or predicativist or whatever standpoint has been codified by some formal system, the way orthodox, classical mathematics is codified by ZFC, the fact that such-and-such is deducible within it is, as discussed in connection with Hilbert's formalism, in effect a finite, combinatorial result and a perfectly correct assertion of orthodox mathematics, a "metatheorem" that can be proved by exhibiting the constructivist or predicativist or whatever proof of the "object theorem" in question.

The accommodation of constructivism and predicativism in this way is admittedly not perfect. Generally, when constructive provability results are announced, it is presumed that constructive provability will be proved constructivistically, by exhibiting a constructive proof of whatever it is; but strictly speaking the bare announcement of the result does not in itself promise this. It might be possible to give a classical, nonconstructive proof that a constructive proof exists without actually producing a constructive proof. To give a rather trivial example, there is automatically a constructive proof of every true numerical identity or inequality, simply by exhibiting computations. Wiles's proof of Fermat's conjecture tells us at once that the following is true:

$$2,345,987^{189,562} + 3,578,041,132^{189,562} \neq 3,987,130,947^{189,562}$$

From its truth we may then at once infer its constructive provability, without our having to exhibit any calculations or even work out any estimates.

There are, in any case, no official censors preventing a group of dissidents from founding a journal of their own, in which as a matter of editorial policy results must be proved according to the group's restrictive standards (or as the case may be, results may be proved making use of the group's preferred additional hypotheses), rather than presented as they would be in a journal enforcing the orthodox standard (as a "metatheorem" about what is provable in some restricted system, or as a theorem with an extra hypothesis explicitly acknowledged in the statement of the result). No one dissident school of thought, however, produces work of sufficient volume at a sufficient pace to keep a journal of high standard following such a policy coming out regularly.

So much, then, for the impact of the "three schools" period of activity in modern philosophy of mathematics, on logic on the one hand and on mathematics on the other hand. And from the point of view of philosophers, how would the period be assessed? The answer is that philosophers have nothing like a unified point of view, even in the loose sense in which logicians and mathematicians may be said to have one. Philosophy is a field in which profound disagreement is the rule rather than the exception. Indeed, as soon as a stable consensus begins to emerge in any branch of philosophy, one ceases to think of it as a branch of philosophy, and begins to think of it as a special science. This is precisely what happened to physics in the seventeenth and eighteenth centuries, psychology in the nineteenth century, and arguably to logic in the twentieth century.

It can only be said that, after the end of the three schools period, philosophy of mathematics became less of a central focus in philosophy. For those who did still work in the field, there was a noticeable trend away from engagement with questions about how mathematics should be practiced, since mathematicians by and large no longer seemed to be philosophically troubled about that. Beyond this, it is always hazardous to say anything about which issues in a given area of philosophy during a given period were the most important ones, since in philosophy claims about the importance of issues are generally about as controversial as any other claims. I will venture to say, however, that the move was largely to questions about how mathematics, as mathematicians practice it, fits into the larger intellectual picture. This is a matter with which not so very many mathematicians concern themselves at all directly—at any rate, not in the same sense and in the same way in which philosophers concern themselves with it—nor is there any obvious reason why they should concern themselves with it.

How mathematics fits in had in a sense been the main question in early modern philosophy of mathematics. Then the puzzle had been how mathematics could proceed, as it appeared to do, by a priori methods, while all other successful branches of science, even those making heavy use of mathematics, as does physics, clearly depended on substantial a posteriori empirical evidence. Post-three-schools philosophy of mathematics has not given up all traditional concern with the apparent a priori character of knowledge in mathematics, but has added a strong concern with the evident "abstract" character of the *objects* of mathematics.

A major transitional figure was W. V. Quine. Raised as a logicist (a student of Russell's collaborator Whitehead), Quine more than anyone is responsible for a turning of philosophers of mathematics towards issues of "ontology," towards questions about the nature of the *objects* of mathematics. Though I myself believe that this reorientation towards ontology was unfortunate, it must be admitted that puzzles about the peculiar nature of mathematical objects do arise quite naturally when one compares mathematics to other scientific disciplines: The difference as to what kinds of questions it makes sense to ask about the objects of the discipline is at least as gross and palpable a difference between mathematics and other sciences as the difference in methodology with which we began this study.

To illustrate the point, cosmologists have long struggled with a problem of "missing mass": The motions of galaxies seem to show the influence of something massive in places where nothing detectable by ordinary telescopes, radio telescopes, or the like is present. One proposed solution has been to attribute mass to neutrinos, elusive but with difficulty detectable subatomic particles originally thought massless. This proposal seems insufficient to solve the problem: Neutrinos may have mass, but not, it seems, enough of it to account for the observed galactic motions. What if some graduate student suggests a new solution, that of attributing mass to *numbers*, supposing them to be mostly concentrated at the cores of galaxies or galactic clusters? Surely it would be instantly recognized that there is a confusion of some sort underlying this proposal?

Numbers just aren't the sort of thing to which it makes sense to attribute mass or charge or whatever, or a spatiotemporal location in remote galaxies or elsewhere. Mathematicians think of some branches of their subject as more concrete and others as more abstract: Number theory is one of the more concrete branches, group theory is more abstract, category theory is even more abstract. But for philosophers, *all* the objects of mathematics are "abstract" and not "concrete," owing to their lack of spatiotemporal location and causal interactions, gravitational, electromagnetic, or other. Their lack even of dates in time distinguishes them not only from material but also from mental entities such as dreams or headaches or *human ideas about* numbers, a point often overlooked by amateur philosophers. And thus they do not seem to fit in with the objects of the other sciences, which all coexist and interact within the same cosmos.

This has led many philosophers to consider their status as problematic, and some to deny their very existence. The label "nominalism," borrowed from a medieval dispute over the status of "universals" such as "redness," has been applied to the view denying the existence of mathematical and other abstract objects. The label "Platonism" has been applied to anti-nominalism (though surely the inference from "Anyone who is a follower of Plato believes in abstract objects" to "Anyone who believes in abstract objects is a follower of Plato" is blatantly fallacious). The issue of nominalism versus Platonism (including as subissues debates among partisans of either side about what is the optimal form or formulation of their side's position) has in many ways been the central issue of post-three-schools philosophy of mathematics, at any rate in Anglophone university philosophy departments.

Quine, to get back to him, for a time joined his colleague Nelson Goodman in embracing a form of nominalism, but soon gave it up. The starting point for his reconsideration of his position was a commitment to taking science seriously, as something more than a useful fiction, combined with pessimism about the prospects for reconstruing or reconstructing it in such a way as to eliminate its dependence on orthodox mathematics. He concluded that there is no way to do modern science without the help of modern mathematics, hence no way to take modern science seriously without taking modern mathematics seriously, which includes taking its existence theorems seriously. (Think of the role of theorems of the form 'There exist solutions to the field equations of general relativity such that . . . " in cosmology, for instance.) But to take seriously the existence theorems is to take seriously the existence of mathematical objects, which are abstract ones. This train of thought is generally called "the indispensability argument." It is difficult to find a statement of all of it all in one place in Quine; see instead Colyvan 2014.

Despite this argument, other philosophers have followed Quine into nominalism and not followed him back out of it. Such philosophers apparently believe that they have an understanding of what "existence" could mean, applied to mathematical objects, that is independent of mathematicians' usual methods of evaluating purported proofs of existence theorems—an understanding by reference to which even what are, by ordinary mathematical standards, the best-established of existence theorems may be judged erroneous or fictitious.

Since we are no longer in a period like that of the three schools, when mathematicians and philosophers alike were concerned with normative questions about how mathematics should be done, nominalist philosophers do not attempt to interfere with the practice of mathematics. Generally speaking, they do not, as Brouwer did, tell mathematicians to change their ways and stop asserting their existence theorems. They merely insist that, inside the philosophy seminar room, one should take back any mathematical existence assertions that may have been made while doing mathematics or applying it in science. And more than a few among the small minority of professional mathematicians who are also amateur philosophers actually agree, and profess "formalist" philosophical theses on Sundays, while making "Platonist" mathematical assertions on weekdays.

The "ontological" period in philosophy of mathematics has run from about 1945 to the present. Some would say that the debate over nominalism and others like it having noticeably run out of steam in recent years, when many of the newer publications only "muddy waters previously clear" (as my colleague and one-time collaborator Gideon Rosen has put it in a review article), we are now in the early years of a fourth period. The new period is supposed to be that of "the philosophy of mathematical practice," which turns back from the generalities of the "ontological" period to specifics about the way mathematics is done, even though there seem to be few "philosophical" controversies as to how mathematics should be done among mathematicians today that are as hotly debated as were the issues of the "three schools" period.[4] Some mathematicians, too, with little patience for technical philosophical discussions inspired by comparison of mathematics with other areas of intellectual life, and so not specifically directed at them, and perhaps also under the influence of a "culture of narcissism," would like to see philosophy of *mathematics* become philosophy of *mathematicians*, and have called for a change of direction.

But whether the "ontological" period is now ending, or still has some years to run, it must be noted that to an extraordinary extent the agenda for philosophers of mathematics for a very long time indeed has been set by just one paradigmatic figure of the period, Paul Benacerraf, and moreover by just two of his papers. One of these, "Mathematical Truth" (1973) revived interest in nominalism at a time when it had flagged a bit, and inspired numerous attempts to respond to Quine's later anti-nominalist

---

[4] There *are*, to be sure, some policy questions, such as those about computer-dependent proofs discussed in Ch. 2, or about the "foundational" ambitions of some category theorists, to be discussed in Ch. 4; but "the philosophy of mathematical practice" is not exclusively, or even very intensively, focused on such issues, which doubtless mathematicians consider themselves competent to settle for themselves, without "help" from outsiders.

position. The other, "What Numbers Could Not Be" (1965) launched for philosophers the issue of "structuralism" in the sense with which we will be concerned here, roughly that brought out at the end of our discussion of Bourbaki. However, to understand what puzzled Benacerraf we must first look a little more closely than I have so far at the "reduction" of mathematics to set theory.

The Peano Postulates

If all mathematics is to be placed in a set-theoretic framework, the natural numbers must be treated set-theoretically. Their treatment in axiomatic set theory is based on a preliminary analysis by Dedekind (1901, part II) which essentially did for number theory what Hilbert was to do for Euclidean geometry: that is, give axioms from which the whole theory, so long as it is considered in isolation, could be developed with strict rigor. Peano's later work, addressed to carrying out parts of that development, can be and generally is viewed as a natural continuation of Dedekind's, and as it happens the axioms proposed by Dedekind have come to be called (despite Peano's own citation of Dedekind) the *Peano postulates*.

These are variations on the five axioms characterizing the order on the natural numbers already listed in this chapter. They provide for an initial or zero natural number 0, and for a next or successor natural number x' for any natural number x, with the specifications that zero is not a successor and that distinct natural numbers have distinct successors, and include the crucial principle of mathematical induction: If the set of natural numbers having a certain property contains 0 and contains x' whenever it contains x, then it contains all natural numbers.

What of addition, multiplication, and exponentiation? Do we not need to include them among our primitives? And what of the various associative and commutative and distributive laws? Do we not need to include them among our postulates? No, addition and the rest can be defined, and associativity and the rest deduced. Dedekind shows that from the Peano postulates it follows that there exists a unique addition operation satisfying the following *recursion equations*:

$$x + 0 = x$$

$$x + y' = (x + y)'$$

The idea of the proof is simply to show by induction that the set of y such that there is a suitable function defined *at least up to y* contains all natural numbers. And similarly for multiplication and exponentiation, as well as several equivalent definitions of less-than.

The recursion equations enable one to derive specific sums, such as 2 + 3 = 5 (where 1 abbreviates 0′, 2 abbreviates 1′ or in other words 0″, and so on). They also enable one to derive general laws. For instance, given x and y the set of z such that the associative law of addition

$$x + (y + z) = (x + y) + z$$

holds for z can be shown to contain 0 using the first recursion equation for + twice, thus:

$$x + (y + 0) = x + y = (x + y) + 0$$

The same set can be shown to contain $z′$ whenever it contains z using the second recursion equation for + thrice, thus:

$$x + (y + z′) = x + (y + z)′ = (x + (y + z))′ =$$
$$((x + y) + z)′ = (x + y) + z′$$

Induction now tells us that the set contains all z, or in other words, that the associative law for addition holds generally. The derivation of the other basic laws of arithmetic is not much harder, provided they are taken up in the right order.

The natural numbers, however, play other roles in mathematics apart from their role as elements of a structure with various arithmetic operations on it. They, or the numerals that name them, "one" and "two" and "three" and so on, are used to count with, and are given as answers to questions of the kind "How many *F*s are there?" or "What is the number of *F*s?" Even if we were only interested in developing number theory on its own, without any regard to external applications, we would need to make provision for this usage, since counting the number of *F*s plays a definite role even in pure number theory. For instance, the Euler totient function ϕ, which makes a very early appearance in number-theory texts, is defined by the condition that $\phi(x)$ is *the number of* numbers y less than x that have no prime factor in common with x. The Möbius function μ is similarly defined in terms of *the number of* prime factors a number has.

However, it turns out that, just as we could define addition and multiplication, and did not need to take them as a primitive, we can likewise define "number of." For the number of Fs is x just in case there is a correspondence between the Fs and the numbers less than x. Moreover, it turns out that, just as we could deduce the basic laws of addition and multiplication, and did not need to take them as postulates, we can likewise deduce the basic laws of number-of. These include, for instance, the law that, if the number of Fs is x and the number of Gs is y, and there are no things that are *both* Fs and Gs, then the number of things that are *either* Fs or Gs is $x + y$. Likewise, the number of pairs consisting of an F and a G will be $x \cdot y$.

Dedekind stops here, for we have by this point got the background presupposed in number-theory texts, and the further development of the theory may be left to number theorists. What the work summarized here shows is that the task of developing number theory within set theory depends simply on being able to define some progression, and prove that it *is* a progression.

The route taken by the axiomatizers of set theory has generally been (i) to define a zero element and a successor operation; (ii) to posit as one of the basic axioms of set theory an *axiom of infinity* according to which there exists a set containing zero and containing the successor of anything it contains; and (iii) to define a natural number to be anything that is an element of *every* set that contains zero and contains the successor of anything it contains. This last step makes induction true essentially by definition.

There have been two main ways this strategy has been implemented, both taking the empty set $\emptyset$, the set with no elements, as the zero element. Zermelo's proposal takes the successor of x to be the singleton set $\{x\}$, the set whose only element is x. Von Neumann's proposal takes the successor of x to be $x \cup \{x\}$, the set whose elements are the elements of x together with x itself. Thus both proposals make $1 = \{0\}$, while Zermelo has $2 = \{1\}$, $3 = \{2\}$, and so on, where Von Neumann has $2 = \{0, 1\}$, $3 = \{0, 1, 2\}$, and so on.

Zermelo's approach is the more straightforward of the two, but Von Neumann's approach has the advantage that it can be extended naturally to the transfinite ordinals: Each ordinal is identified with the set of its predecessors, so that the first transfinite ordinal ω is identified with the set $\{0, 1, 2, \dots\}$ of finite ordinals or natural numbers, and the next transfinite ordinal $\omega + 1$ with the set $\{0, 1, 2, \dots, \omega\}$, and so on. In specialist works in set theory, intended only for other set theorists, Von Neumann's identification is generally tacitly assumed.

Benacerraf's Puzzle

Even these days, when the rudiments of set theory are introduced in the elementary grades, children are still taught to count and do arithmetic before exposure to set theory. Benacerraf, however, imagines a child being brought up to learn set theory first, and then about the natural numbers through the set-theoretic treatment of them. Or rather, he imagines two children, one being brought up on the Zermelodic approach, and the other on the Neumannian. In their encounters with other, more normally raised children and with adults, even including mathematicians, or at least "core" mathematicians—mathematicians minus set theorists and other specialists in "foundations"—nothing will give away the fact that they have had such unusual upbringings. They will agree with everyone else on basic sums such as 2 + 3 = 5 and basic laws such as associativity of addition, and the further consequences of such laws as far as they have studied them, as well as in their answers to "how many?" questions, on which the applications of arithmetic depend.

This just about covers the range of topics likely to come up when the children discuss the natural numbers—except with each other. Between themselves, however, they will discover differences. The Neumannian will hold that 0 *is* an element of 2, but the Zermelodist will hold that 0 is *not* an element of 2, and they will find it very hard to locate any considerations that might help themselves decide which of them is right. And the conclusion Benacerraf draws is essentially that there is nothing to be right about: There is no right "identification" of numbers as sets, because numbers simply are not sets; or at least, nothing in mathematical practice justifies assuming them to be sets.

Well, it would have been extremely implausible for either Zermelo or Von Neumann to have claimed to have discovered what natural numbers really had been all along, without Archimedes or Newton or Gauss ever noticing, and neither makes such a fantastic claim. But Benacerraf has another target, Frege—the most relevant parts of his work are in the Benacerraf and Putnam anthology—who *does* seem in at least some of his formulations to be making a claim of this kind for his rival analysis of number. (Insofar as he is concerned with Frege, Benacerraf's paper overlaps one by Charles Parsons from the same year, 1965.)

Frege had his own theory of collections, different enough from mainline set theory that one generally uses "class" and "member" in place of "set" and "element" for the relevant notions of collection and belonging. Cantor's theory of sets has a "from-the-bottom-up" character: Since, as already mentioned, one can only gather together what are there to be gathered, sets are conceived of as arising by metaphorically "gathering" *pre-existing* elements. Later elaborations emphasize that one can then go on to form sets of sets by metaphorically "gathering" sets, and then sets of sets of sets and more; but one never reaches a comprehensive set of all sets. Frege's theory has a "from-the-top-down" character by contrast. One starts with the class of all and everything, and arrives at other classes by dividing it up, separating things that have some property from things that don't. And Frege really did at least sometimes write as if he thought he had discovered what numbers really were, and that they were certain of his classes.

In addition to Zermelo's and Von Neumann's and Frege's definitions or analyses of number, one must consider the position of Cantor in the pre-axiomatic period of the development of set theory. A common background to Cantor or Frege is provided by the tendency of ordinary language and commonsense thought to move quickly from noticing two items "have something in common" to positing some *thing* that they have in common. In more technical language, the move is from recognizing some notion of *equivalence* applicable to some old sort of item to recognizing a new sort of item, *equivalence types* of things of the old sort. The equivalence type is the new sort of item that two items of the old sort that are equivalent in whatever respect thereby have in common.

Frege's favorite example should make plain what is meant: Lines (the things of the old sort) that are parallel (equivalent in the relevant notion of equivalence) thereby have in common their *directions* (the equivalence types). This trivial-seeming example is actually of some importance, since the initially mysterious "points at infinity" of projective geometry, at which parallel lines meet, may be explained as nothing more than directions in this sense. Often the only word in ordinary language for the kind of equivalence in question will be one that presupposes the types. Thus figures that are similar in the technical geometric sense thereby have in common their *shapes*, but the only non-technical word for "similar" in the relevant sense seems to be "like-shaped," though one can introduce a pretentious Latinate synonym "equiform."

The properties and relations that it makes sense to apply to equivalence types are those that correspond to properties or relations of the things they are types of that are *invariant* in the sense of continuing to hold even if equivalents are substituted for equivalents. Thus if a first line is perpendicular to a second, any line parallel to the first will be perpendicular to any line parallel to the second; and this invariance allows us to call directions "perpendicular" in a transferred sense, if they are the directions of lines that are perpendicular. Similarly, a figure is *convex* if given any two points in the figure, the whole line segment connecting them lies entirely inside the figure, as with a circle but not a crescent; otherwise the figure is *concave*. Since any figure like-shaped to a convex (respectively, concave) figure is itself convex (respectively, concave), we may apply "convex" and "concave" in transferred senses to shapes.

In general, properties pertaining to spatial location are not invariant: Parallel lines and similar figures are located in different places, and so one cannot speak of the spatial location of directions or shapes. This is one feature that makes them paradigmatically "abstract" as that epithet is used by philosophers. Indeed, an alternate word for "equivalence types" is "abstracts," and the label "abstraction" is often used for the process of positing equivalence types of previously recognized entities with respect to one or another relation of equivalence.

Two structures that are isomorphic are said to have thereby in common their isomorphism type. Thus isomorphism types are the equivalence types or abstracts associated with the relation of being isomorphic. Some mathematicians may regard talk of "isomorphism types" as merely a manner of speaking, not regarding "structures" in the sense of isomorphism types as the kind of objects one would want to develop a mathematical theory about, the way one does with "structures" in the sense of spaces or number systems or whatever. But if we do take isomorphism types seriously, they give us a way to conceptualize cardinal and ordinal numbers.

First note that while a structure in general consists of a set plus something extra, and an isomorphism is a correspondence between sets that respects the something extra, a bare set may be considered a degenerate case of a structure, where isomorphism amounts to nothing but correspondence. In this special or degenerate case it is customary to say "equinumerous" rather than "isomorphic," since pre-theoretically, the existence of a correspondence between two sets is the criterion for the elements of the one being as numerous as the elements of the other. What two bare sets that are isomorphic, which is to say, two sets that are equinumerous, thereby have in common is their *cardinal number*. Or at least, that is more or less how Cantor conceived of cardinal numbers; and he conceived of ordinal numbers as essentially isomorphism types of well-ordered sets.[5]

Frege begins with a conception of numbers as, essentially, the equinumerosity types of classes of things, and the natural numbers as the equinumerosity types of *finite* classes of things. However, he takes the further step of, essentially, identifying an equivalence type with the class of all items of that type. Thus the direction of a line becomes the class of all lines parallel to the given line, and the number two becomes the class of all two-membered sets or unordered pairs or couples. Frege feared that without this identification there would be no basis for our firm naive belief that, whatever the number two may be, it isn't Julius Caesar! Commentators have scratched their heads over why Frege thought we could tell for sure that Caesar is not a class, when we supposedly cannot be sure that he is not a number, unless we have not been told that numbers are classes. (The "Julius Caesar problem" has been endlessly discussed in the literature on Frege.)

It may be admitted that Frege's is a somewhat less artificial-seeming identification than the expedients resorted to by Zermelo or Von Neumann within orthodox set theory, which were needed precisely because the two-membered sets or unordered pairs or couples are too many to form a set. And in general the representation of equivalence types as set-theoretic objects has to be artificial for the same reason. (The general method was, indeed, developed only in the 1960s, by Dana Scott, and it depends on the rather technical "axiom of foundation.")

---

[5] This is a somewhat sanitized version of Cantor's view. Cantor himself spoke as if cardinal and ordinal numbers were *mental* entities. Starting with some ordered set of objects, by a mental act of inattention to their nature, to everything about them except how they are ordered, one arrives at their *order type*, which if they are well ordered is an ordinal number. Then by a second mental act of further inattention even to how they are ordered, and hence to everything about them except their being distinct from each other, one arrives at their cardinal number. Old-fashioned notations such as $|X|$ for the ordinal number and $||X||$ for the cardinal number reflect Cantor's talk of single and double "abstraction" or selective inattention. Frege polemicized against this sort of "psychologism," pointing out that the predicates it makes sense to apply to a number and those which it makes sense to apply to a human idea are entirely different. Cantor's psychologistic or mentalistic or conceptualistic or idealistic account seems indeed absurd even on Cantor's own showing, since he maintains that there is an absolute infinity of ordinal and of cardinal numbers, which is surely incompatible with each number being a product of some human mental act.

But Benacerraf shares the skepticism most readers tend to feel about the claim that this is what the number two really was all along.[6] And one reason for skepticism is that "multiple reductions," or other candidate identifications, are available on Frege's approach to collections as much as on that of axiomatic set theory. One could, for example, identify the number two with the class of all classes having *no more than* two members, or the class of all classes having *at least* two members (and there are other possibilities whose description would take us too deeply into Fregeanism or Russellianism).

We may grant Benacerraf this point: It cannot be claimed that numbers all along were Fregean classes any more than it can be claimed that all along they were sets. But what about the more Cantorian conception of natural numbers as equinumerosity types of finite sets *without* the Fregean identification of equivalence types with classes? That may appeal to those who think of the natural numbers primarily as finite *cardinals*. But alongside the cardinal numerals "one" and "two" and so on, one has the ordinal numerals "first" and "second" and so on, used to answer questions of the kind "In which place in the A-order does object *a* come?" And so one may think of natural numbers not only as finite cardinals, but as finite *ordinals*. And that suggests a different identification, that of natural numbers with isomorphism types of finite orders.

So it seems that, even adopting a naive, uncritical attitude towards the notion of equivalence types, there are still "multiple reductions" available. Frege (1983) gave cardinal notions priority over ordinal notions, and Cantor the reverse, and developmental psychologists such as the famous Jean Piaget have weighed in with views (fairly complicated ones not easily

---

[6]   Russell, who made the same identification of the number two with the class of all couples, differed from Frege by explicitly disclaiming any ambition to tell us what the number two really was all along. In his *Introduction to Mathematical Philosophy* (1919, 18) he writes: "So far we have not suggested anything in the slightest degree paradoxical. But when we come to the actual definition of numbers we cannot avoid what must at first sight seem a paradox, though this impression will soon wear off. We naturally think that the class of couples (for example) is something different from the number 2. But there is no doubt about the class of couples; it is indubitable and not difficult to define, whereas the number 2, in any other sense, is a metaphysical entity about which we can never feel sure that it exists or that we have tracked it down. It is therefore more prudent to content ourselves with the class of couples, which we are sure of, than to hunt for a problematical number 2 which must always remain elusive." From the standpoint of axiomatic set theory, the class of couples is *not* something "we can be sure of": There are too many couples for there to be a set of all of them, and this is a key respect in which Frege's conception of class differs from Cantor's conception of set.

summarized) about which has priority in a child's acquisition of number-concepts (see Beth and Piaget 1974). But many are left with the feeling that the natural numbers are not *more* cardinals than ordinals or vice versa, in which case, cardinals and ordinals being distinct things, natural numbers cannot be either the one or the other.

There remains the possibility that natural numbers are objects *sui generis*, "of their own kind," not to be identified with any objects not introduced as numbers, be these sets, or classes, or equivalence types conceived of as something different from sets or classes, or anything else. But Benacerraf, having eliminated so many candidates for what sorts of objects natural numbers could be, moves on quickly to suggest (though not to endorse unequivocally) the conclusion that they are not *objects* at all.

Structuralism: Hardheaded and Mystical

Robert Lowell, in a well-known line from the *Life Studies* poem "For George Santayana," sums up that philosopher's "Catholic atheism" or "cultural Catholicism" in the dictum "There is no God, and Mary is His mother." The Benacerrafic view might be summarized in a similarly paradoxical formulation: There are no natural numbers, and infinitely many of them are prime. In fact, Benacerraf's seminal paper closes with a formulation of just this type. A less puzzling formulation of his view would distinguish two levels of discourse, a higher, "meta" level talking *about* mathematical practice, and a lower, "object" level talking *within* mathematical practice.

At the higher level, we say that there is no such thing as the natural number system N, and explain this to mean that, though there are many progressions, no single one of them has an exclusive right to be styled "*the* natural numbers" with the definite article. Each progression has elements in its initial, next-to-initial, next-to-next-to-initial, and so on, positions, but since no progression has an exclusive right to be called N, so no elements occupying such positions have an exclusive right to be called *the* number 0 or 1 or 2 or whatever.

At the lower level, we do speak of "the" natural number system N, and "the" natural numbers 0, 1, 2, and so on, but this is really only a manner of speaking; we say nothing about N except what could be said about *any* progression, and nothing about 0, 1, 2, and so on, except what could be said about the initial, next-to-initial, next-to-next-to-initial, and so on, elements of such a progression. What we can say includes everything that follows from the Peano postulates, and we have seen that this includes everything that would be said in ordinary mathematical practice. In particular, it includes Euclid's theorem that there exist infinitely many primes. The Santayanesque formulation, "There are no natural numbers and infinitely many of them are prime," obviously not meant to be taken literally, is an instance of something akin to the figure of speech known as zeugma.[7] Its oddity results from its dropping from the "meta" to the "object" level in mid-sentence, despite the anaphora of the pronoun "them" in the second half with antecedent "numbers" in the first half.

The Benacerrafic version of structuralism is an instance of what Dummett called *hardheaded* structuralism, and contrasted with the *mystical* structuralism he thought he saw in Dedekind. To the question whether we may still speak of *the* natural number system, hardheaded structuralism answers "No." As I mentioned earlier, there are also structuralists who answer "Yes, but." These are the mystics. More fully, they answer "Yes, but the structure in question is a very peculiar one."

The mystical view posits a unique structure that is indeed *the* one and only natural number system N. It is supposed to be a progression *with no other properties beyond that of being a progression*, along with whatever follows from that (which we have seen in the discussion of the Peano postulates is, after all, quite a lot). The claim seems paradoxical: Isn't having no other properties than being a progression *itself* a property, and moreover a very strange and distinctive one that presumably no other progression has? In this connection, Parsons calls the lack of other properties a "metaproperty" rather than a "property."

On this view, there is also an item that is indeed *the* one and only natural number 0: It is the initial element in the structure that is the one and only natural number system N; and similarly for 1, 2, and the rest. None of the natural numbers 0, 1, 2, and so on, has any properties beyond that of occupying the place it does (as the initial, next-to-initial, next-to-next-to-initial, or whatever element) in the structure N, and whatever follows from

---

[7] This figure or trope is traditionally illustrated by a couplet from Pope containing an apostrophe to Queen Anne: "Here Thou, great Anna! whom three Realms obey, / Dost sometimes Counsel take—and sometimes Tea." In "sometimes take counsel and sometimes take tea" there would merely be use of the same word "take," first in one sense, then in another. In the contracted or elliptical version there is only one occurrence of the key word, and it must be taken in a double sense.

these: They have the "metaproperty" of having only "structural" properties. Like N itself they are "property-deficient." For it is rather obvious that
the individual natural numbers have to be property-deficient if the structure N is to be so. If the number 2 had, for instance, the property of having conquered Gaul, the structure N would have the property of having a
conqueror of Gaul in its next-to-next-to-initial place. Likewise, it seems,
if the individual numbers are to be property-deficient, the structure must
be so as well.

Now there is uncontroversially *something* that there is just one of connected with progressions, namely, their common isomorphism type.
The isomorphism type is, however, not itself generally thought of as a
structure, and in particular, not as a structure of the kind of which it is
the isomorphism type. The isomorphism type is like a "universal" on an
Aristotelian conception, in which redness is exemplified by red things, but
is not itself one of them, being abstract and hence colorless. The mystic's
Natural Number Structure is like a Form on a Platonic conception, for
which Triangularity is itself triangular. Mystical structuralism is "mystical" in just the sense that Platonism (in an historically serious sense, not
the silly contemporary sense of anything-but-nominalism) is mystical.

It may be helpful to compare the disagreement between actual hardheaded and mystical structuralists over the "univalent" theory of numbers
by comparing it to a disagreement between hypothetical hardheaded and
mystical interpreters of the "multivalent" theory of groups. The first fact
to be explained is why, in so much of group theory, group theorists discuss groups at length without ever mentioning the individual natures of
their elements, the properties they have apart from those pertaining to
the relationships to each other within the group (relationships such as z's
being the product of y and x, or y's being the inverse of x, or x's being the
identity).

The mystical explanation of this first fact is that the group theorists
are speaking about extraordinary, property-deficient objects, "abstract
groups" whose elements *have* no individual natures. And this may seem
the simplest and most natural explanation so long as we consider the first
fact only. A rival hardheaded explanation is not lacking, however. It would
be that the group theorists are making general statements about *all* groups,
regardless of the individual natures of their elements, and since they are
making no assumptions about those individual natures, they never mention those individual natures. One is not *forced* to interpret an absence of

assertions as an assertion of absence, or the treatment of groups "in the abstract" as treatment of "abstract groups" as mystics conceive of them.

The hardheaded will also point to a second fact, that the results of pure group theory get applied to groups whose elements do have individual natures, in some cases as the permutations of roots of equations, in other cases as geometric transformations of some space, and in yet other cases something else again. It is in this way that group theory connects with other branches of mathematics such as the theory of equations, various geometries, and further subjects. According to the hardheaded, the results of pure group theory are thus applicable because they are general, and apply to groups of all kinds, and not just special property-deficient "abstract groups." The theory would be a pretty idle luxury if it were just about *those*. And now that we are considering the second as well as the first fact, the hardheaded view may appear the simplest and most natural explanation of what is going on. A rival mystical explanation is not lacking, however. It would be that the results about abstract groups are applicable because every "concrete" group is isomorphic to an abstract group. And so we may seem to be at a stand-off.

One contrast between the hardheaded and mystical views of number theory is immediate. A mystic can say that the number 2 has no location in space or date in time, no causes and no effects, no mass and no charge, and so on. For to have a location in space or date in time or causes or effects or mass or charge would be to have a non-structural property, while the mystic's view is that the number 2 has none. And this perhaps seems the most natural thing to say if confronted with the graduate student with the new speculation about the missing-mass problem.

The hardheaded structuralist, while speaking at the "lower" or "object" level, cannot say that the number 2 lacks spatial location, since that isn't true of the next-to-next-to-initial element of absolutely *any* progression. (Given any progression, we can obtain another by replacing its next-to-next-to-initial element by Julius Caesar, for instance.) What the hardheaded structuralist *can* say is that, whatever specific location is considered, we would not be justified in attributing to the number 2 the property of being located at *that* location. (For whatever the location specified may be, taking any progression we can get another by replacing whatever element is in its next-to-next-to-initial place by one that is not at the specified location.) That may be enough to answer the graduate student; but a franker expression of the hardheaded view would be one that can be given only speaking at the "higher" or "meta" level: Numbers cannot be ascribed mass, not because they are massless, but because they are nonexistent.

But the issue of whether the number 2 is located in physical space, and if so where, is a metaphysical one, and is not one that arises in mathematical practice. The differences between the two species of structuralists will not be apparent from what they say when doing mathematics, but only from what they say when doing philosophy. This means that the hardheaded and the mystical structuralist, with their rival interpretations of talk of "the natural numbers" or "the number two"—one taking it as a *façon de parler* not to be understood as being literally about any one specific progression and its next-to-next-to-initial element, and the other taking it to be about specific but property-deficient objects—are in a situation actually quite closely analogous to that of the Neumannian who thinks that $2 = \{0, 1\}$ and the Zermelodist who thinks that $2 = \{1\}$. It is a bit surprising that no successor of Benacerraf has come forward to assert on this ground that both sides are wrong.

Descriptive versus Revisionary

The Dummettian terms in which I have so far been describing the varieties of structuralism, "hardheaded" and "mystical," have pejorative connotations for some, and are not the terms preferred by the actual adherents of the views in question. There is no single, agreed alternative terminology, but Geoffrey Hellman has suggested "*sui generis* structuralism" for the mystical view, and Charles Parsons "eliminative structuralism" for the hardheaded one.[8] The label "eliminative" is appropriate insofar as Benacerraf's suggestion "eliminates" the natural numbers in the sense of denying (when speaking at the "meta" level) that there are such things.

---

[8] The reader who wishes to pursue the issue will find that there are four book-length works mainly or largely devoted to advocating structuralism: Hellman 1993; Resnik 1997; Shapiro 1997; Chihara 2003; and about half of Parsons 2008 is devoted to the issue as well. Hellman and Chihara are clearly on the hardheaded side, Resnik and Shapiro on the mystical side. Shapiro, borrowing terminology from the medieval debate over universals, uses "*ante rem* structuralism" for his own mystical or *sui generis* view, on which structures exist "before the things" that exemplify them, and "*in re* structuralism" for the opposing hardheaded or eliminativist view, on which structures exist only "in the things" that exemplify them.

In fact, those structuralists who are eliminativists are generally nominalists, and would "eliminate" *all* mathematical objects. So far, I have been discussing primarily structuralist accounts of the natural numbers, but a general elimination of mathematical objects would obviously require extending structuralism across the board. Indeed, *sui generis* structuralists, too, now often write as if they advocated a view about mathematical objects across the board parallel to the view of natural numbers that I have just been describing. But before undertaking to evaluate the two varieties of structuralism as across-the-board theses, I need to say something about the standpoint from which I will be evaluating them.

The prominent Oxford philosopher P. F. Strawson introduced a distinction that will be useful in this connection, between what he called *descriptive* metaphysics and what he called *revisionary* metaphysics. Descriptive metaphysics is concerned to trace the broadest categories of common-sense and scientific thought, without pretending that philosophy is able to outdo common sense and science and produce better ones. Revisionary metaphysics, by contrast, rejects the ordinary categories and proposes others; it may grant that for everyday use, outside the philosophy seminar room, the ordinary categories may be good enough, but it insists that on entering the philosophy seminar room one must acknowledge that they are at best useful fictions.

Nominalists who profess to be pursuing descriptive metaphysics make *hermeneutic* claims to the effect that, properly understood, the statements asserted by mathematicians and scientists that on the surface appear to imply or presuppose the existence of abstract objects really deep down mean something different and nominalistically acceptable, or else to the effect that the mathematicians and scientists who appear to assert such statements are not really asserting them, but performing some other "speech act." The first type of claim is called *content* hermeneuticism, and the second *attitude* hermeneuticism. From either it follows that in saying there are no functions, for instance, the nominalist philosopher is not really contradicting anything that a mathematical physicists who says "There are solutions to the field equations of general relativity such that . . ." seriously means to assert.

Such hermeneutic nominalists contrast with *revolutionary* nominalists, frankly revisionary metaphysicians who acknowledge they are proposing to reject existing commonsense and scientific theories except as useful fictions. One may perhaps speak outside the philosophy seminar room as if one believes these fictions, keeping one's fingers crossed, but on entering the philosophy seminar room one must take it all back. Inside the philosophy seminar room, the lazier revolutionaries will offer no new theory beyond the assertion *that* the fictions prevailing among non-philosophers are useful, while the more ambitious revolutionaries will offer some new theory attempting to explain *why* these fictions are as useful as they are. This new theory may, however, consist of nothing more than the words of the old theory with some new and previously unintended meanings attached.

The same descriptive/revisionary or hermeneutic/revolutionary division found in metaphysics and ontology is found also in epistemology, the theory of knowledge. Here the *naturalized* epistemologist is one who has become a citizen of the scientific community, and judges by its standards, while the *alienated* epistemologist remains a foreigner to that community, judging its doings from an external standpoint. Many nominalists, under the influence of Benacerraf's "Mathematical Truth," hold there to be some serious difficulty about how one could come to have a *justified* belief in anything implying or presupposing the existence of abstract objects. All revolutionary nominalists who take such a line are, whether they acknowledge it or not, engaging in alienated epistemology. For they have tacitly rejected the possibility that the actual historical process by which mathematicians and physicists have come believe that there are, say, solutions to the field equations of general relativity with such-and-such characteristics, might be a process of formation of a *justified* belief. They are therefore implicitly presupposing some extraordinary philosophical standards of justification, by which ordinary scientific standards of justification whose historical application has led mathematicians and physicists to their present views, including ordinary mathematical criteria for evaluating existence proofs, can be found wanting, except as routes to useful fictions.

In this chapter I will set aside all hermeneutic versions of nominalism, which I have discussed at some length elsewhere,[9] and likewise set

---

[9] See Burgess and Rosen 1997 and Rosen and Burgess 2005. The one point from those previous discussions it seems important to reiterate here is that a philosophical argument, whether derived from Benacerraf or from elsewhere, to the effect that it cannot be justifiably believed that abstract entities exist, can never supply the hermeneutic nominalist with more than a reason for *hoping* that some nominalistically acceptable reinterpretation of mathematicians' and scientists' words or attitudes can be found; such an argument can never in itself provide evidence for *believing* that any proposed reinterpretation is correct.

aside all issues of alienated epistemology and revisionary metaphysics, to consider things only from the standpoint of scientifically informed common sense, claiming no access to any supposed sources of higher and better philosophical standards of judgment. In this sense, I am concerned with "mathematical practice" and will be evaluating mystical and hardheaded, or *sui generis* and eliminative, structuralism only insofar as they are put forward as interpretations of mathematical practice, and not as proposals to replace existing mathematical practice by something else, or to retain it outside the philosophy seminar room, but disown it inside.

The reader should understand that this is not necessarily the standpoint from which leading structuralists are advocating structuralism. But while the standpoint of mathematical practice may not be the structuralists' *preferred* standpoint, they cannot claim that it is an *irrelevant* standpoint from which to evaluate their position, since there is a very conspicuous appeal to mathematical practice early on in Benacerraf's motivating discussion: It is because the difference between the account on which $2 = \{1\}$ and the account on which $2 = \{0,1\}$ *makes no difference to (core) mathematical practice* that one is supposed to doubt that either identification can be better or truer than the other.

Nominalism, Vacuity, Modality

Let us return now to the comparison of the mystical or *sui generis* version of structuralism with the hardheaded structuralist or eliminativist. The latter seems to face a problem that the former does not. If the statements of number theorists are to be reconstrued as or replaced by generalizations about *all* progressions, rather than singular statements about some one progression, *the* natural number system, these reconstruals or replacements will be vacuous unless there is *at least one* progression. The relevant structure must be *exemplified* somehow. This is no problem for the mystic, who does not construe number-theoretic statements as generalizations about a class of structures one has to worry may turn out to be empty, but posits a *sui generis*, property-deficient Natural Number Structure, the Platonic Form of a Progression, for number theory to be about.

Some will say that the advantages of mystical over eliminativist structuralism here are (in a famous phrase of Russell) "the advantages of theft over honest toil," but be that as it may, they are undeniably advantages.

For the eliminativist's worry is a serious one, especially if, as is virtually always the case, the eliminativist is a nominalist. For in seeking an exemplification, these nominalists, since they deny the existence of abstract, mathematical objects, will have to seek it among concrete, physical (or mental) objects. But for there to be a progression of physical objects (or mental objects) there will have to be *infinitely many* such objects, which is a dubious assumption about the physical world (and presumably a plain false one about the mental world).

The problem becomes more acute if, as again is virtually always the case, the eliminativist wants to claim that the statements of analysis, also, are to be taken as generalizations, this time about complete ordered fields. Exemplification now seems to require an *uncountable* infinity of physical objects. The problem becomes much more acute still if the eliminativist wants to claim that the statements of set theory, besides, are to be taken as generalizations, this time about set-theoretic universes, also called cumulative hierarchies. Exemplification then seems to require an *absolute* infinity of physical objects, more than any transfinite cardinal number of them!

Now as Hilbert emphasized at the beginning of his lecture "On the Infinite" (see Hilbert 1983), physics provides us with little or no reason to suppose that there are infinitely many *physical* objects. To be sure, physics generally uses real number coordinates to represent points of space, or rather, point-instants of space-time, and thus models space-time as something continuous, with an uncountable infinity of locations available. At least one nominalist, Hartry Field, has been willing to accept point-instants of space-time as "physical" entities in some sense (see Field 1980), thus providing for uncountably many such entities. The philosopher Philip Ehrlich has suggested that in modeling space-time it might conceivably be useful to use, not real numbers, but the *surreal* numbers of Conway 1976, of which there are an absolute infinity. If one combined Field's with Ehrlich's suggestions, one would get as many "physical" entities as any eliminative project could have use for.

But as philosophers of mathematics and science have pointed out—the *locus classicus* is in the work of Penelope Maddy (1997, *passim*, and elsewhere)—this line of thought seems to take the mathematical modeling of physical space more seriously than physicists themselves take it, so that the whole idea of getting vast infinities of "physical" entities in this way involves a kind of confusion over what "physical" can really mean. And so the eliminativist, who denies there are any abstract, mathematical enti-ties to provide exemplifications if there fail to be enough concrete, physi-cal entities to do so, seems to face the problem that for the real-number structure, let alone for the set-theoretic-universe structure, there will be no exemplifications at all, rendering the theories involved vacuous.

Some have wished at this point to posit that, though there may not *actu-ally* be enough entities, at least there *could possibly have been* enough. That would be no help if one continued to take mathematical theories as only being about such exemplifying structures as there actually are. Hence an accompanying suggestion is that number theory, for instance, should be taken as being not about what *actually is* true in all the progressions there actually are, but as about what *necessarily must be* true in any progressions there actually are or possibly could have been. Since possibility, necessity, and related notions are traditionally called *modal* concepts, this approach is called the modal strategy. Hellman, its best-known advocate, wants to consider "modal structuralism" one of *three* varieties of structuralism, but Parsons rightly observes that it is merely one strategy for eliminativism.[10]

Modal structuralism faces a number of difficulties, some technical, some philosophical, concerning what must be assumed about "necessity" and "possibility" to implement the strategy, and about how plausible such assumptions are. I will not pursue these issues here, because by the time one comes to modal structuralism one is obviously pretty well into the ter-ritory of revisionism rather than descriptivism.[11]

Indeed, I would contend that one has crossed over from description to revision at least as soon as one generalizes structuralism from a view about the natural numbers and the elements of a few other structures con-sidered in "univalent" theories, to a general thesis about all mathematical

[10] Hellman (2001) has given a short account of his way of counting structuralisms. He counts bourbachique structuralism (which I have suggested is nowadays almost a trivial tru-ism) as one variety, Shapiro's *sui generis* (less politely, mystical) structuralism as a second, and his own modal structuralism as a third (forgetting about eliminativist strategies that might appeal to, say, point-instants of space-time rather than modality). In a footnote he suggests that there may be a fourth variety, connected with category theory, but defers that topic, as I will defer it here.

[11] This is clear enough from some of the reasons Hellman (2001) offers for thinking bour-bachique structuralism is not structuralism enough. Some of the issues not to be treated here, mostly technical ones, are dealt with in Burgess and Rosen 1997, which critically surveys various nominalist projects, modal structuralism among them. The first half of Parsons 2008 has dealt with others of those issues, mostly more purely philosophical ones. Parsons offers an extended critique of eliminativism and of the role of modality in eliminativist projects.

objects, enunciated in such formulations as "Mathematical objects are featureless points in structures; there is nothing more to them than what follows from their occupying the places they do in their structures."[12]

Such a formulation may be offered by a mystical or *sui generis* structuralist as an account of how things are, or by a hardheaded or eliminativist structuralist as an account of how things would have to be if we took mathematics "at face value" and did not regard it, as the eliminativist structuralist holds we ought to regard it, as a mere *façon de parler*. Either way, I submit, this kind of formulation is untrue to the diverse ways in which diverse kinds of mathematical objects are treated in mathematical practice, which to repeat is the only standpoint from which I wish to consider the issues here.

"Featureless Points"

Let us consider the mathematical objects that would be encountered in, say, a first course in number theory. Do these seem to be treated by number theorists as featureless points in structures? We may begin with the central objects of study, the natural numbers.

The starting point for structuralist reflection was the observation that if the natural numbers have intrinsic natures apart from their positions in the natural number system, these are never mentioned. Everything that is said about them is compatible with a structuralist interpretation, and indeed with either of two structuralist interpretations: the mystical interpretation that they are property-deficient specific objects, and the hardheaded interpretation that there are no such things and apparent talk of

---

[12] And structuralists do say such things, though the formulation just given is not that of any single author, but a pastiche of several. Specifically, my formulation runs together two to be found at the beginning of Parsons 1990, one by Parsons himself as part of the body of the article, one by Resnik in a passage that is quoted with apparent approval. The former reads as follows: "By the 'structuralist view' of mathematical objects, I mean the view that reference to mathematical objects is always in the context of some background structure, and that the objects involved have no more to them than can be expressed in terms of the basic relations of the structure." And the latter goes thus: "In mathematics, I claim, we do not have objects with an 'internal' composition arranged in structures, we have only structures. The objects of mathematics, that is, the entities which our mathematical constants and quantifiers denote, are structureless points or positions in structures. As positions in structures, they have no identity or features outside of a structure." Parsons 1990 significantly qualifies the bold views expressed in these formulations in the course of developing them.

them is merely a way of expressing generalizations about arbitrary systems conforming to certain axiomatic assumptions.

But equally everything that is said about them is compatible with the non-structuralist interpretation that they do have a specific nature that there is simply no occasion to mention, at least until one turns from pure theory to applications. They might, for all that is said about them, be finite cardinals conceived as equinumerosity types. For starting from that conception, one could take the basic laws relating operations such as addition on numbers to operations such as union on collections as *definitions* of the arithmetic operations. Basic laws about arithmetic operations, such as the associativity of addition, would follow then from corresponding laws about set-theoretic operations, such as the associativity of union. (This was essentially the Frege-Russell logicist approach.) The recursion equations for the arithmetic operations, the basic properties of order, the Peano postulates about zero and successor, would all be derived from set or class theory, reversing the order of ideas found in developments that *start* from the Peano postulates. The point is that even a first course in number theory starts with all the background apparatus already set up, to be taken for granted in what follows, with no clue in what follows as to how it was originally arrived at; and consequently it seems there is no basis on which to choose between a hardheaded structuralist, a mystical structuralist, and a non-structuralist account.

And all this pertains to the system $\mathbb{N}$ of natural numbers, the most favorable case for structuralist interpretations! Other mathematical objects are soon introduced for which structuralist interpretations may be more difficult to arrive at. Almost immediately one introduces negative numbers, to expand $\mathbb{N}$ to the (ordered) "ring" $\mathbb{Z}$ of integers, and then the fractions, to expand $\mathbb{Z}$ to the (ordered) "field" $\mathbb{Q}$ of rational numbers. The expansion of $\mathbb{Q}$ to the full "complete ordered field" $\mathbb{R}$ of real numbers, and this in turn to its "algebraic closure" $\mathbb{C}$, the complex numbers, may not be undertaken at once (or at all, in a first course), but individual irrational numbers such as the quadratic surds (the square-root $\sqrt{2}$, the golden ratio $(1 + \sqrt{5})/2$, and the like) and individual imaginary numbers such as the Gaussian integers (the imaginary unit i, the factors $1 + i$ and $1 - i$ whose product is 2, and the like) make an early appearance.

Right away the idea that each number has a *unique* structure as its home is in trouble, since 2 seems equally at home in all of them, unless one wants to insist that the natural number 2 must be distinguished from the integer +2, the rational number 2/1, the real number 2.000 . . ., the complex number 2.000 + 0.000 . . . *i*, and so on (as admittedly is done in some symbolic computation programs), and that mathematicians' habitually writing of them as if they were all the same is a case of "abuse of language" (which is admittedly an extremely common phenomenon in mathematics). Some structuralists, however, prefer at this stage to elaborate or qualify their view, to allow that one structure may be a substructure of another, or to distinguish "basic" structures from "derived" structures.

The objects of number-theoretic investigation include also *sets* of natural numbers or integers or whatever, and *functions* from and to the natural numbers or integers or whatever. And I do not mean merely that one speaks of certain particular sets such as the powers of two or the primes, or certain particular functions such as the operations + and ·, or the Euler φ and Möbius μ mentioned earlier. I mean that one makes what logicians call "second-order" assertions "quantifying over" sets and functions: One asserts generalizations about *all* number-sets or numerical functions. An example involving sets right at the beginning of the subject would be the *well-ordering* or *least-number* principle: The usual order on the natural numbers is a well-ordering; every nonempty set of natural numbers has a least element. An example involving functions would be the *Möbius inversion formula*, which makes an early appearance in elementary textbooks.

Sets and functions are not treated as featureless positions in structures, lacking intrinsic nature. Clearly, Cantor did not regard sets as "featureless points" connected by a relation that for unspecified reasons was called "elementhood," and was for unspecified reasons assumed to obey certain laws. Rather, a set was treated as having a specific nature, as an *unum* formed *e pluribus,* a single thing formed from a plurality of things, the elements to which the set is intrinsically related. What laws are obeyed by sets so understood was a matter for investigation based on this prior understanding. Something like Cantor's understanding seems still to prevail when one speaks of sets of natural numbers in number theory. Likewise, a function is treated, not as a "featureless point," but as something whose nature is to take arguments or inputs and give values or outputs.

To be sure, insofar as it is up for grabs what "the" natural numbers are, it is up for grabs what sets of them are or functions from and to them are: Number-sets and numerical functions are tied to the natural numbers in such a way that, if one changes one's conception of the latter, one must change one's conception of the former. In this sense, sets and functions are tied to structures even if they are not themselves positions in structures, and accommodating versions of structuralism may be content with their having this much of a "structuralist" nature, if not more.

The real difficulties for a structuralist account of mathematics, if one is ambitious enough to wish to apply it to literally *all* the objects studied by mathematicians, ironically concern "structures" in Bourbaki's sense: the groups, rings, and fields that pervade number theory in its modern form, even at the basic, introductory level, to say nothing of more sophisticated developments. The student will make the acquaintance of the ring $\mathbb{Z}[i]$ of Gaussian integers and the rings $\mathbb{Z}_m$ of integers modulo m, perhaps before being given the general definition of "ring." And I mean not only that mention is made of particular such structures, but also that universal theorems about whole classes of such structures are proved,[13] beginning with the result that one can always do division (except by zero) in $\mathbb{Z}_m$ if and only if m is a prime. Such rings are hardly treated as "featureless" or "structureless." They are treated as collections of things with "additional structure" in the form of two operations, one written additively, the other multiplicatively.

The structuralist may have no way of handling such structures other than that of taking the desperate route of (i) embracing after all the set-theoretic reduction of mathematics, which turns the rings in question, along with everything else, into items in the universe of set theory; and then (ii) giving a structuralist account of *set theory* of one kind or another. Set theory would have to be interpreted as being not about a specific kind of thing whose nature is to be a collection of other things, but rather about a mystical structure known as the set-theoretic universe, in which "sets" are mere featureless points, or else as dealing in generalizations about *all* structures of an appropriate kind. But as already mentioned in connection with number-sets, there simply is no grounds in mathematical practice, at least not at the level of beginning number theory, for taking set theory as anything other than what it appears to be, a theory of things whose nature is to be collections of other things. In short, whatever may be the attractions of across-the-board or global structuralism as revisionary metaphysics (and whatever may be its appeal to alienated epistemologists), it is not plausible

---

[13]   All this the author can say with confidence as a former student, counselor, and parent of students in the Ross Program, a summer math camp for high-school students.

as an account of mathematical practice. Or so it seems for the moment; we will revisit the issue briefly later, when there will be something more positive to say.

Indifference in Practice

Though the preceding remarks have been critical of structuralism, and most of the remarks to follow will be as well, it should be emphasized that Benacerraf, and following him the various structuralist philosophers of mathematics, have performed a real service by pointing out to philosophers a real phenomenon, a kind of *indifference* on the part of working mathematicians, namely, an indifference to exactly how one got to the point from which their own investigations begin. The number theorist, even in a first course, wants to take certain background material for granted, and is generally indifferent as to where the background notions and results came from. Moreover, Benacerraf is right that included in the scope of this indifference is indifference as to the *identity* of the natural numbers, to which of the many isomorphic progressions is "the" natural number system.

Structuralism errs in generalizing too far from these initial correct observations—too far, and in the wrong direction. As I begin to enlarge upon this claim, let me first of all note that it would be wrong to say that mathematicians are quite generally indifferent to the differences between isomorphic structures. (I do not say that any structuralist has been incautious enough to claim the contrary, but only that they do not emphasize the point as much as I think it should be emphasized.) A simple example can be drawn from group theory, and something like it may be found in the first pages of any book or chapter on that subject.

In an old-fashioned three-way light switch, supposing the switch is initially in the off position, a rotation of 90° will light a first filament, a rotation of a further 90° to 180° will light a second as well, rotation of a further 90° to 270° will turn off the first but leave the second on, while rotation of a further 90° to a full circle of 360°, which comes to the same thing as the original position of 0°, will turn off the second as well. If we denote rotations or turns of the switch through 0°, 90°, 180°, and 270° respectively as *a*, *b*, *c*, and *d*, and consider performing first one rotation and then another as a kind of "multiplication," then we will get the multiplication operation

Table 3.1. Cyclic group of order 4

	A	b	c	d
a	a	b	c	d
b	b	c	d	a
c	c	d	a	b
d	d	a	b	c

Table 3.2. Noncyclic group of order 4

	a	b	c	D
a	a	b	c	D
b	b	a	d	C
c	c	d	a	B
d	d	c	b	A

indicated in the adjoining Table 3.1. In a two-switch set-up, controlling a light from either of two places, there are again four operations that can be considered: (*a*) leaving the switches alone; (*b*) flipping the first switch; (*c*) flipping the second switch; or (*d*) flipping both switches. In this case, we obtain the multiplication table in the adjoining Table 3.2.

One may find mathematicians saying, "There are only two groups with four elements, the *cyclic* one, and the *noncyclic* one." Yet one may also find mathematicians saying, "The group of permutations of four elements has four distinct subgroups that are noncyclic groups with four elements." Evidently, these two statements cannot both be meant literally. The one meant figuratively is the first, and the figure of speech involved is ellipsis. The statement is elliptical for "There are only two groups with four elements, *up to isomorphism*," a locution meaning "There are only two *isomorphism types* of groups with four elements," or more fully, "There are two non-isomorphic groups with four elements such that every group with four elements is isomorphic to one or the other of them," the two being those shown in the tables.

Galois theory, alluded to in the first part of this study, is one area where distinguishing multiple isomorphic smaller groups among the subgroups of a large group is very common. Group representation theory, which has not been alluded to previously, and which it would take us too far afield to go into, may be cited as a branch of mathematics that is in a sense concerned with *nothing but* differences between isomorphic groups. Fields medalist Terence Tao (2013, 9) writing for students about the role of abstraction in mathematics offers one example in the words, "Algebra often ignores how objects are constructed or what they are, but focuses instead on what operations can be performed on them, and what identities these operations enjoy. (This is in contrast to representation theory, which takes the complementary point of view.)" Structuralist philosophy of mathematics seems to focus too exclusively on what is just one of two complementary (and interacting) approaches. No unqualified, general claim that mathematicians are unconcerned with differences between isomorphic structures is tenable.

More importantly, perhaps, in our present context, there are many areas where mathematicians *are* indifferent, but where their indifference cannot be described as indifference *between two isomorphic structures*, since there are no such structures involved. The case cited by Benacerraf, that of indifference between a set-theoretic definition of 2 as [{∅}] and of 2 as {∅, {∅}}, is only one of countless cases of indifference connected with set-theoretic reductions. The case Benacerraf cites does involve, or can be construed as involving, some kind of indifference between isomorphic structures; but just this feature makes the case an atypical instance of the general phenomenon it exemplifies.

Many kinds of indifference can be seen in the way in which mathematicians speak of structures. For instance, a topological group consists of a set X, a multiplication operation $\cdot$ such as would give us a group, and a family of open sets O such as would give us a topological space, subject to the axioms of group theory and topology plus one more connecting $\cdot$ to O. In a full-fledged set-theoretic reduction, a topological group would have to be identified with a set of some kind, an item in the set-theoretic universe. Supposing X and $\cdot$ and O have been so identified, the topological group can be identified with the ordered triple $(X, \cdot, O)$, where an ordered triple (x, y, z) is identified with an ordered pair whose second item is itself an ordered pair, and an ordered pair (x, y) can be identified with the set $\{\{x\}, \{x, y\}\}$.

But there are several arbitrary choices here. The topological group could equally well have been identified with the ordered triple $(X, O, \cdot)$, the ordered triple with the ordered pair $((x, y), z)$, and the ordered pair with the set $\{\{x, ∅\}, \{y, \{∅\}\}\}$. In practice, nothing a working mathematician says will give away what definition of ordered pair or ordered triple is involved, and even if a mathematician in one lecture says that a topological group is a triple $(X, \cdot, O)$, the same mathematician in the next lecture may very well be found saying that a topological group is a triple $(X, O, \cdot)$.[14] We might expect a book to choose one option and stick to it, but if two books make opposite choices, no one is going to say that they are addressing different topics: "M's book is about groups $(X, \cdot)$ with topological structure O added, while N's book is about topological spaces (X, O) with group structure $\cdot$ added."

Here we have a kind of indifference to the identity of an object that is not a matter of stripping the object of any properties but those that follow from its occupying the position it does in some background structure. For topological groups are themselves background structures for many developments, but are far from being featureless, or mere points in some still larger, still deeper background structure. The topological group example is at least still "ontological," indifference concerning an *object*, but mathematician's indifference extends to "ideology" as well, and in particular, to questions about what the definition of this or that notion is.

Most interesting notions admit of several provably equivalent characterizations, and if one book takes characterization A as the definition and characterization B as a theorem about the notion defined, while another book does the opposite, mathematicians will generally not treat the difference as important mathematically (and *perhaps* not even pedagogically). For instance, what is the definition of the number e, the basis for the natural logarithms? I am not asking here what real numbers are, but rather, assuming we're satisfied about real numbers in general, what I'm asking about is which of them is e. There are two well-known characterizations, one as a limit, the other as a series:

$$e = \lim_{n \to \infty} (1 + n^{-1})^n$$
$$e = 1/0! + 1/1! + 1/2! + 1/3! + \cdots$$

---

[14] This sort of point is by now a commonplace. I recall first hearing it made by my dissertation supervisor Jack Silver in the early 1970s. Parsons mentions yet another form of indifference in this area: A function is treated in set-theoretic reductions as a set of (argument, value) ordered pairs, but they could equally be taken to be sets of (value, argument) ordered pairs, and mathematicians generally are content to specify a function by specifying what is its value for a given argument, without displaying it as any kind of set of ordered pairs. It is noteworthy that Bourbaki never explicitly identifies a topological group with an ordered triple, and at least in the 1st edn of the volume on set theory, did not offer any set-theoretic identification of ordered pairs, but took pairing as a primitive notion.

It makes absolutely no difference whatsoever to subsequent developments whether one takes the first as one's definition and derives the second from that definition as a theorem, or whether one does the reverse.

The phenomenon of indifference is very widespread, though philosophers' characteristic preoccupation with "ontology" has perhaps blinded them to its full extent. The phenomenon is sufficiently widespread that even if one wholly accepted a structuralist account in the case of indifference between two isomorphic structures, some account of indifference of other kinds would be called for. How much more must such an account seem needed if one harbors doubts about structuralism! A clue to an account can be found in the mathematical literature itself, and from a period before talk of "structure" became fashionable in any quarter. To this matter I will next turn.

Hardy's Principle

Discussion so far has focused on the case of the natural numbers, with occasional mention of the real numbers along with them. But the case of the real numbers deserves a closer look. In Greek mathematics (I mean the rigorous side of it), "number" always meant "natural number," and these began, not with 0 as they do for us, and really not even with 1, but with 2. There were no negative numbers. Nor were there any fractions, so far as theoretical mathematics as opposed to business arithmetic is concerned. Instead of speaking of rational numbers such as $^2/_3$, Euclid or Archimedes or Apollonius will speak of a ratio of 2:3. To say that such ratios were not regarded as numbers is to say that one did not speak of adding or multiplying them. Still less were there irrational real numbers. Rather, there were ratios $a:b$ of lengths (or line segments) and other magnitudes, some of which, such as the ratio of diagonal to side in a square, were not equal to any ratio of whole numbers.

In Omar Khayyam, one finds such ratios treated as numbers. For instance, Euclid knew how, given ratios of lengths $a:b$ and $a:c$, to construct a length d such that the proportion $d:c::b:a$ would hold, and Khayyam considers $a:d$ to be the *product* of $a:b$ and $a:c$. So long as one only "multiplied" geometric magnitudes, the product of two lengths would yield an area, and the product of this with a third length would yield a volume, and then one could not go further. Ratios, however, can be multiplied indefinitely.

Such is the chief advantage of Khayyam's conception. One still finds essentially the same conception in Newton's *Universal Arithmetick*.

This geometric approach can be made completely rigorous, according to the highest standards of rigor, and this is in fact done in Hilbert's *Foundations of Geometry*; but by that time mathematicians were firmly committed to "arithmetization," the program already alluded to in the first half of this study, of making analysis independent of Euclidean geometry. The best-known contribution to that project was Dedekind's (1901, part I). As in his work on natural numbers, in his work on real numbers he isolated the crucial properties of the numbers in question. For the reals, what is key is the kind of "completeness" of the real numbers discussed earlier in connection with the rigorization of the calculus. This feature distinguishes the reals from other ordered fields beginning with the rational numbers.

In the background was the Greek theory of proportion, attributed to Eudoxus, and available to us in Euclid, book 5. If there is a *common measure* for lengths a and b, which is to say, a shorter length some whole number m of copies of which exactly fills up a, and some other whole number n of copies of which exactly fills up b, then the two lengths are called *commensurable*, and their ratio $a:b$ is taken to be equal to $m:n$. Proportionality $a:b::c:d$ or equality of the ratios $a:b$ and $c:d$ for two other lengths c and d then holds if the ratio $c:d$ is also equal to $m:n$. The problem facing Eudoxus was to define proportionality in the case of incommensurable lengths.

Roughly speaking, the idea was this. If dividing b into n equal pieces and laying off m such pieces along a falls short of the end of a, then we may say that $a:b$ is greater than $m:n$, whereas if laying off m such pieces along a overshoots the end of a, then we may say that $a:b$ is less than $m:n$. Eudoxus' definition amounts to this, that $a:b$ and $c:d$ are equal if and only if for every m and n, either $m:n$ is less than both, or $m:n$ is greater than both. For us moderns, for whom ratios of whole numbers are rational numbers, and ratios of lengths are real numbers, this says that two real numbers are the same if and only if for every rational number, either it is less than both or it is greater than both. In other words, a real number is completely determined by its place among the rational numbers, by the "cut" that it makes among rational numbers, dividing them into a lower class, consisting of all rationals that are less than the given real, and an upper class, consisting of all those that are not less. The lower class will never have a greatest element, while the upper class will have a least element just in case the real number is itself rational, in which case it is the least element of its own upper class.

Dedekind's idea—one of those things that seems obvious after it is pointed out, however unobvious it was before—was simply to drop the geometric background here, and just take real numbers to be items introduced by cuts, divisions of the rationals into two classes, both nonempty, with every member of one class, called the lower, less than every member of the other class, called the upper, and with no greatest element in the lower class. He spoke of real numbers as being determined by or corresponding to cuts, but it has become customary to speak of the real number as simply *being* the cut. Strictly speaking, since a rational number is clearly something different from the cut it determines, this would make the real number .5000 ... corresponding to the rational number $1/2$ something distinct from the rational number itself; but the construction can be adjusted to avoid this result.

It remains to define the algebraic operations on real numbers (in a way independent of the geometric constructions that defined them for Newton), and to verify the basic laws of algebra. The definition of addition is more or less obvious: The lower cut of the sum of two real numbers will consist of all sums of two rational numbers, one belonging to the lower cut of the first real number, and the other to the other. The verification of associativity of addition for real numbers will invoke associativity of addition for rational numbers. And similarly for multiplication and for the commutative and distributive laws and more. The verifications are tedious to write out in full, but essentially routine, involving no further major ideas. In the end, we get all the laws of a so-called complete ordered field, and the development can be capped off by a proof that any two complete ordered fields are isomorphic, the counterpart for analysis of the proof that any two progressions are isomorphic in arithmetic.

By the beginning of the twentieth century, the rigorous approach to calculus and analysis was beginning to find its way into textbooks for undergraduate students of mathematics. Britain, though it had become less isolated from the Continent over the course of the nineteenth century, lagged a little behind. I have already mentioned the textbook of G. H. Hardy (the analytic number theorist we met earlier as the "discoverer" of Ramanujan), *A Course of Pure Mathematics*, intended for first-year students at Cambridge. By the second edition (1914), Hardy was including an account of the Dedekind construction just reviewed.

Hardy's account of the construction is followed by an interesting remark:

The reader should observe ... that no particular logical importance is to be attached to the precise form of the definition of "real number" that we have adopted. We defined a "real number" as being . . . a pair of classes. We might equally well have defined it as being the lower, or the upper class; indeed, it would be easy to define an infinity of classes of entities each of which would possess the properties of the class of real numbers. What is essential in mathematics is that its symbols should be capable of *some* interpretation; generally they are capable of *many*, and then, so far as mathematics is concerned, it does not matter which we adopt. (Hardy 1914, 15)

We may call this last remark *Hardy's principle*.

Hardy could have mentioned that, in addition to the various minor variations on the Dedekind construction that he cites, there was a totally different construction, also leading to a complete ordered field, and so isomorphic in its results to the Dedekind construction, due to Cantor. As with all stages of the nineteenth-century rigorization process, there are generalizations, and Dedekind and Cantor both provide methods of introducing a "completion" that in the case of the rationals yields the reals, but that can also be applied in other contexts. (The *p*-adic numbers, briefly alluded to earlier, are another example of a completion.)

Hardy goes on to say the following:

Bertrand Russell has said that "mathematics is the science in which we do not know what we are talking about, and do not care whether what we say about it is true", a remark which is expressed in the form of a paradox, but which in reality embodies a number of important truths. It would take too long to analyze the meaning of Russell's epigram in detail, but one at any rate of its implications is this, that the symbols of mathematics are capable of varying interpretations, and that we are in general at liberty to adopt whichever we prefer. (Hardy 1914, 15–16)

In line with Hardy and with his take on Russell,[15] different textbooks introduce the real numbers in different ways, and while reviewers may find one approach *pedagogically* superior to another, no one considers any definition *mathematically* superior to any other, so long as that other still suffices for deducing the basic properties of the real numbers, those of a complete ordered field. Two analysts who wish to collaborate do not need

---

[15]   What Hardy quotes is from the work reprinted as Russell 1956. Its opening line ("The nineteenth century, which prided itself . . .") has already been quoted here.

to check whether they were both taught the same definition of "real number," as conceivably two algebraists may have to check whether they are both using the same definition of "ring" (since on some usages the definition includes having a multiplicative identity, while for others it does not). For it is only the properties of a complete ordered field that will be used in their collaboration, and never the definition of "real number."

Thus while a textbook may offer a definition of the real field as composed of pairs of sets of rationals *à la Dedekind*, or alternatively as classes of sequences of rationals *à la Cantor*, and on the basis of its definition prove the theorem that the real field is a complete ordered field, only the theorem and not the definition has the status of being generally accepted by the mathematical community and used in subsequent mathematical work. The situation exactly parallels that with Zermelo's and Von Neumann's definitions of natural numbers. Once the construction is done, and a number system with the right properties set up, one never looks back (or one does only if one is looking to do something analogous, as in the case just mentioned of the *p*-adic numbers).

Hardy, though perfectly aware of the phenomenon, not being a philosopher preoccupied with issues of "ontology," does not draw any conclusions about the nature of the real numbers, in the way that the mystical structuralist would draw the conclusion that they are property-deficient objects, or the hardheaded structuralist would draw the conclusion that they do not exist. Hardy's principle pertains not to being but to *meaning*.

The mainstream of academic philosophy in the English-speaking world is commonly called "analytic" and is reputed among many nonphilosophers to have made a "linguistic turn," approaching all questions through analysis of the language in which they are posed. If that characterization were actually correct, as perhaps it was at some early stage of the development of the analytic tradition, one would not find present-day Anglophone philosophers asking "What is the number two?" but rather "What does the numeral 'two' denote?"

As soon as the question is put this way one sees that the presupposition that the word is used to denote something is not self-evidently true, and that the question should really be "How is the numeral 'two' used?" As soon as the question is put this way one sees that the presupposition that there is a single use for the expression, the same across all times and places, is not self-evidently true, and that the question should really be, "How is the numeral 'two' used, say, in contemporary, professional mathematics?," leaving room for other questions about historical, lay, and other uses. To ask how an expression is used is nearly the same as asking, or at any rate is very closely related to asking, what the word means, and so one would expect the post-linguistic-turn, analytic philosopher to be raising the question of meaning. But it is rather Hardy, the mathematician, who raises the issue of meaning (in the quotation), while the philosophers, preoccupied with "ontology," are still asking what (or whether) the number two *is*. There is a certain irony here.

Hardy's principle makes no distinction between "ontological" and "non-ontological" cases, between the definition of the symbol "$\mathbb{R}$" for the real number system and the definition of the symbol "e" for the base of the natural logarithms. In either case, according to Hardy, what is important is that it should be possible to give some meaning to the symbols which would make what the mathematician says when using them true, while what may be a matter of indifference is just what meaning one chooses, what definition one imposes. One need not think that a mathematician such as Hardy, while working in analytic number theory, has any one definition of "$\mathbb{R}$" or of "e" specifically in mind.

For the phenomenon that has drawn the attention of structuralist philosophers, Hardy's principle suggests an explanation independent of any contentious specifically ontological assumptions. I believe it is possible to go a step further, and give an explanation of sorts for Hardy's principle itself, as derivative from the nature of rigor as it is observed in present-day mathematics. That is what I will try to do next.

Codification versus Codifiability

Euclid covers many topics in ancient mathematics, but never intended to cover the whole terrain, just "the elements." Bourbaki was even further from having an intention to cover all the vaster terrain of modern mathematics. The initial intentions of the group were comparatively modest, and though they soon became more ambitious, to the end (or to the present, for the Bourbaki seminar still meets) certain areas were excluded from consideration beyond rudiments that might be needed for other areas. Set theory is a conspicuous example, but far from the only one. An important founder of the group has written a survey (Dieudonné 1982) which is quite open about the group's hierarchy of interests, in

STRUCTURE AND STRUCTURALISM 155

which the bottom rank (D) includes set theory, but also classical analysis. Nonetheless, the Bourbaki series allows one to imagine what an all-encompassing codification might look like.

Now what I want to suggest is that the way the ideal of rigor operates is by requiring that each new mathematical paper should be written as if it were a longer or shorter chapter or section within such a compilation. And what would that mean? Just this, that the author of a new piece of work is responsible for the logical cogency of the definitions of any new notions from old ones, those found in previous chapters and sections, and for the logical cogency of the deduction of new results from old ones, again to be found in previous chapters and sections, together with the new definitions. Of course, by "previous chapters and sections" I mean the earlier literature, metaphorically thought of as if constituting such parts of the same grand encyclopedia.

And what does this leave out? Just this, any responsibility for how any notions taken from old work were defined in terms of still older notions and so on until ultimately we get a definition in terms of primitives, or for how any results taken from old work were deduced from still older work and so on until ultimately we get a deduction from postulates. One may, of course, want to go into such background, if one wants to generalize some earlier definition or deduction. I am not supposing the hypothetical encyclopedia would be bound by the actual principle of the Bourbaki group never to do this, but always to state everything from the beginning in the most general possible form that could ever be needed in subsequent work. That principle requires superhuman foresight concerning what may eventually be wanted.[16]

The principle of never looking inside previous results, treating them as black boxes, is in fact adhered to by Euclid, though not at all the principle of stating things in maximal generality the first time around. I believe that, in one sense, this principle suffices to cover all the kinds of indifference of which note has been taken. For it implies that, if there are different routes by which the material (notions and results) on which the author immediately relies could have been obtained from first principles (primitives and postulates), the author can be indifferent to any choice among them.

---

[16] Adherence to the principle by the actual Bourbaki group brought down on themselves not only complaints on pedagogical grounds, but also the wrath of the category theorists, whose subject had not been invented at the time the Bourbaki project was launched, and for whose perspective no allowance was therefore initially made. But of that, more later.

Any actual codification would have to make choices among rival definitions, both of the non-ontological kind we saw with the rival definitions
of *e*, and of the more ontological kind we saw with the rival definitions of
R, and of other kinds as well. In any actual codification, one would have
to choose for each important notion a single characterization in terms of
more basic notions to take as its definition, and for each important type
of mathematical entity a single construction out of more basic entities to
take as their definition. In the actual Bourbaki encyclopedia, as far as it
goes, any number of such choices are indeed made. A hypothetic extension to an ideal encyclopedia of all of mathematics would have to make
many more. But what mathematical rigor requires, I am suggesting, is not
actual codification, but only potential codifiability. It must be possible to
view new work *as if* it were a chapter in a codification, but it need not really
be such a chapter, and may remain indifferent to all the many more or less
arbitrary or conventional choices that would have to have been made on
the way to the immediately preceding chapter.

In another sense, the principle that only potential codifiability, not actual
codification, matters does not explain *any* of the observed indifference that
has been under discussion. The principle tells us that any aspect of old work
not needed or useful for new work can be disregarded, but it does not tell us
which aspects of old work these are likely to be. In many cases, including the
cases that first drew the attention of philosophers, and led to philosophical
structuralism in philosophy of mathematics, what one can be indifferent
to are different ways of constructing a structure with desired properties,
leading to distinct but isomorphic results. But I have already observed that
the slogan that differences between isomorphic structures can be neglected
goes in some respects too far and in other respects not far enough.

I have no rival general principle to suggest, and am suspicious that there
are no really general principles, but at most rules of thumb with limited
ranges of applicability, with the principle that differences between isomorphic structures are *likely candidates* for neglect being only the most
conspicuous. There may in the end be no substitute, if one wants to
know what matters and what not, to reading the actual mathematics, or
if that is impracticable, at least semi-popular accounts thereof by actual
mathematicians.

One corollary follows from the suggested account of the scope and limits
of indifference, namely that the mathematician working on advanced topics probably never has to consider what the first principles *are*. And indeed,

in my own experience with core mathematicians, I would say that for many of them the most important thing about foundations is that they should not have to think about them. I say this not exclusively on the basis of personal communications, but also on the basis of several decades' observation of mathematicians' more or less public reactions to results obtained by set theorists and other logicians. I have in mind, in particular, the reaction, perennially disappointing to logicians, to *undecidability* results.

Gödel showed that for any theory T in which one can do a modicum of mathematics, there will be questions formulable in the language of T that are undecidable on the basis of T, in the sense that neither an affirmative nor a negative answer to the question can be deduced from T.[17] If T is ZFC, then this means that the question is not decidable by ordinary mathematical means. The statement that T is consistent is one example. It can be expressed as a less "metamathematical," more mathematical-looking example using a coding of symbols as numbers of a kind now familiar from electronic transmission of messages as sequences of zeros and ones, the digits of a binary numeral. Work of Gödel, then of Martin Davis and Hilary Putnam and Julia Robinson, and finally of Yuri Matiyasevich, solving what had been the tenth problem on a famous list of 23 put forward by Hilbert, shows that the undecidable statement can always be cast in the form of the statement that a certain Diophantine equation (polynomial equation in several variables for which only integer solutions are to be admitted) has no solution.

But the equations are not of a kind whose solution would be high on the list of priorities of number theorists. The first "natural" example was the continuum hypothesis (CH), which had been Hilbert's first problem. Gödel showed in the 1930s that, if ZFC is consistent, then it cannot refute CH, while Paul Cohen showed in the 1960s that if ZFC is consistent, then it cannot prove CH either. But CH is very much a question of higher set theory, remote from the concerns of the great bulk of working mathematicians.

---

[17] In the case of theories with "second-order" axioms, such as the Peano postulates or the theory of complete ordered fields, that are "univalent" or categorical in the sense that all their models are isomorphic, some answer will be a *consequence* of the theory in the sense of being true in all its models. But the difference between first-order and second-order logic is that for the latter there are no complete proof- or deduction-procedures, and so there will still be questions for which no answer is *deducible*. (There are no categorical *first-order* theories, except ones all of whose models are finite.)

The methods of Gödel and especially Cohen, further developed by Robert Solovay and others, have produced a long list of "natural" examples, one of the more striking of which was Saharon Shelah's proof of the undecidability of a famous conjecture of J. H. C. Whitehead in the theory of Abelian groups. Harvey Friedman has for decades been producing ever more concrete-looking combinatorial examples of undecidability as well (specializing in ones that *can* be decided on the basis of "large cardinal" hypotheses going beyond the usual axioms of ZFC).

The response of mathematicians has pretty uniformly been to conclude that either (i) the question shown as undecidable was posed in too general a form, and that the really interesting question is a more restricted one to which the undecidability result would not apply; or (ii) that the branch of mathematics to which the question belongs is somehow "peripheral" and not "core," in the way that set theory itself, once one gets past the rudiments, is seen as a peripheral and not a core field. Probably nothing short of a demonstration (which no one expects) that the twin primes conjecture, say, is undecidable, or the Riemann hypothesis, would be enough to get mathematicians to sit up and take notice of logic and set theory in the way that logicians and set theorists may think they should.

A recent attempt to describe the working mathematician's attitude in a more contemporary idiom has recently been made by the combinatoricist Gil Kalai, who writes in a blog (2008):

While mathematical logic and set theory indeed make up the language spanning all fields of mathematics, mathematicians rarely speak it. To borrow notions from computers, basic mathematical logic can be regarded as the "machine language" for mathematicians who usually use much higher languages and who do not worry about "compilation." (Compilation is the process of translating a high programming language into machine language.)

Here "not worrying about compilation" seems to be a very apt contemporary metaphor for the phenomenon of indifference I have been discussing.

Ironically, it may be largely because core mathematicians by and large don't perceive set-theoretic undecidability results as affecting their own work that they acquiesce in set theory having a semi-official status as the "foundation" of mathematics. A "foundation" that required one to be constantly taking note of it would, it seems, be much less desirable than one that can in practice be allowed to repose largely unread in chapter 1 of a virtual encyclopedia now up to chapter 1001.

4

Structure and Foundations

Foundations and Architecture

The foregoing discussion brings us to the question of the "foundational" claims of certain category theorists, who would like their own work to appear in the opening chapter of the virtual encyclopedia. But before addressing the contentious and confusing issue of the "foundational" pretensions of category theory, I had better say something about the original *non*-foundational role of category theory, and before getting into any specifics about that, I had better say something about Bourbaki's contrast between "foundations" and "architecture."

The metaphor is highly suggestive. In a building, the foundations are mostly subsoil and out of sight, and are not what one shows off when showing off the edifice. They are the province of the structural engineer, not the architect. For Bourbaki, "architecture" refers to what begins to be brought up only at the end of the opening volume of the series, after the earlier "foundational" sections: the general theory of structures.

On one level, the material there is supposed to help implement the policy of doing things at the highest level of generality the first time around, by giving a definition of isomorphism that will be applicable to all the different kinds of structures to be considered in later volumes, so that one will not have to give separate definitions of isomorphism of groups, isomorphism of rings, homeomorphism of topological spaces, and so forth, as one comes to consider those different types of structures in detail. On another level, the choice of examples emphasized, groups and topological spaces, with ordered sets as a poor third, suggests ideas about the organization of mathematics. Groups typify algebraic structures with operations, while topological spaces are the most basic geometrical structures. This in itself does not take us very far beyond the traditional division of mathematics into an algebraic side and a geometric side, with analysis as a third. An organizational scheme that would take one much further was emerging, while Bourbaki was still at work on some of the earlier volumes of the series in the 1940s, in the form of category theory—emerging from a circle of ideas very much present, as category theory itself was not, in early Bourbaki.

The first appearance of the notion of category was in a modest, auxiliary role in a 1945 paper of Samuel Eilenberg and Saunders Mac Lane (né MacLane). Rigorous mathematics, consisting of rigorously proved theorems stated in terms of rigorously defined notions, is surrounded by a range of conjectures and sometimes whole "programs," involving some claims that can at present be rigorously stated but have not yet been rigorously proved, and others that cannot as yet even be rigorously stated. Adherence to the ideal of rigor does not consist in doing without such material, which can play an important role in guiding research, but simply in distinguishing it sharply from what has actually been rigorously established, in a way that might not seem important to physicists or engineers. Eilenberg and Mac Lane were concerned with giving a rigorous definition of one as-yet-not-rigorously defined notion, that of "naturalness" in a certain sense.

To say what it was about requires going just a little beyond freshman-year calculus, perhaps, but may be worth doing, since the example is so often referred to in later semi-popular writing about category theory, and is bound to be encountered by the reader exploring the literature. Abstracting from Heaviside's original vector calculus, a *vector space V* over the real numbers consists of objects, called *vectors*, on which one can perform two operations, addition of two vectors u and v to form the sum $u + v$, and multiplication by a real number a to form the multiple au of a vector. Hence, we can form more generally *linear combinations* of vectors

$$au + bv + cw + \cdots$$

The operations must conform to certain laws, such as associativity of addition. A set of vectors is *linearly independent* if no element can be expressed as a linear combination of other elements, and a linearly independent set is a *basis* for the vector space if every vector can be expressed as a linear combination of its elements. It can be shown that any two bases for the same vector space are equinumerous, and the cardinal number of a basis is called the *dimension* of the vector space. Any two vector spaces of the same dimension are isomorphic. We will be interested only in the finite dimensional case. A *linear functional* on the vector space is a function F assigning real numbers to vectors, in such a way that

$$F(au + bv + cw + \cdots) = aF(u) + bF(v) + cF(w) + \cdots$$

If F and G are linear functionals, so is their sum, and likewise any multiple, where sum and multiple are defined by

$$(F + G)(u) = F(u) + G(u)$$

$$(aF)(u) = aF(u)$$

Indeed, the linear functionals on V themselves form a vector space (in other words, the associative and other laws are obeyed), called the *dual* V^* of V. And the dual in turn has its own dual, the *double dual* of V^{**} of V. Now if the dimension is finite, say for illustrative purposes three, then V and V^* are isomorphic: If u, v, w form a basis for V, then a basis for V^* is given by

$$F(au + bv + cw) = a$$

$$G(au + bv + cw) = b$$

$$H(au + bv + cw) = c$$

It immediately follows that the double dual V^{**} is likewise isomorphic.

But there is a difference between the two cases, which it was Eilenberg and Mac Lane's first task to characterize. There is a "natural" way to define directly an isomorphism between V and V^{**}. To each element u of V, we need to associate a linear functional of linear functionals, which is to say, function assigning to each linear functional F a number. We can do so by assigning to u the function that assigns to F the number $F(u)$. By contrast, defining an isomorphism between V and V^* in the way shown requires picking a basis u, v, w, and we get different isomorphisms from different choices of basis. There seems to be no "natural" way to pick a single isomorphism that deserves to be considered special.

The key to the analysis of "naturalness" proposed by Eilenberg and Mac Lane was a certain kind of uniformity, the possibility of assigning isomorphisms not case by case, but for all vector spaces at once, by means of what they called a *functor*. Such a thing is a function of a certain kind, and the domain on which it is defined is the *category* of all vector spaces.

Thus, originally, categories were introduced for the sake of functors, and functors for the sake of defining naturalness. The categories soon came to have a life of their own, to begin with mainly in solo work of Mac Lane, and in the opinion of many give a better "architectural" account of, or "organizational" scheme for, mathematics than does Bourbaki's general notion of structure.

In a subject like group theory there are "axioms," but not more than the first page of the development of the theory consists in making deductions from the axioms. The axioms serve to define a kind of structure, the models of the axioms, but then almost immediately one turns from establishing consequences of the axioms true in all models to looking at connections between different models. This is characteristic of all "multivalent" theories. The most important connections are established by the kind of partial structure-preserving maps known as *homomorphisms*.

The main architectural insights of Mac Lane I would describe as being (i) that in every branch where there are structures, there are not only structures but "morphisms" of one kind or another connecting them to each other, so that what one is really studying is not just a range of structures, but a set-up with structures and morphisms; and (ii) once the right notion of "morphism" has been found, all the relationships and connections of importance between the relevant structures can be defined in terms of morphisms and composition of morphisms, or replaced by notions thus definable that can serve all the purposes of the original. At this point (ii) is a little cryptic; what is meant will, I hope, gradually become clearer. Sticking with (i) for the moment, the notion of a *category* is in the first instance just a label for set-ups of this kind. (Actually, this is what is called a *concrete* category; a more abstract theory comes later.)

Thus in group theory one has to deal with not just with groups but groups and homomorphisms, in topology one has to deal with not just topological spaces, but topological spaces and continuous functions from one to another.[1] To return to formulation (ii), what is essential to

---

[1] The notion of continuity here, generalizing the notion from analysis, is that a map f from a space X to a space Y is *continuous* if for every open set V in Y the *inverse image* set U of all x in X such that $f(x)$ is in V is one of the open sets in X. The V and U here are generalizing the role of an open interval $(f(x) - \varepsilon, f(x) + \varepsilon)$ around the value and an open interval $(x - \delta, x + \delta)$ around the argument in the Weierstrass approach. The partial structure-preserving aspect here goes *backwards* from Y to X, whereas with group homomorphism the partial structure-preserving aspect goes *forwards*. With a homomorphism f from G to H, if $a \cdot b = c$

the category-theoretic approach is that *all* connections are made by means of morphisms: All relationships of interest must either find a definition in terms of morphisms, or be replaced. This principle may be called the *Categorical Imperative*.

The Categorical Imperative

The situation can be illustrated by the notion of the *product* of two groups G and H. The elements of the product are pairs (g, h) consisting of an element of G and an element of H, and the group operation is given by

$$(g_1, h_1) \cdot (g_2, h_2) = (g_1 \cdot g_2, h_1 \cdot h_2)$$

where the g_i are multiplied in the sense of the operation on G and the h_i are multiplied in the sense of the operation on H. Along with the product group $G \otimes H$ we get two homomorphisms, p from $G \otimes H$ to G, taking (g, h) to g and q from $G \otimes H$ to H, taking (g, h) to h. Together $K = G \otimes H$ and p and q have the *universal property* that for any group L and any homomorphisms r from L to G and s from L to H there is a unique homomorphism t from L to K such that r is the composition of t and p (the result of first applying t then applying r to the result), and s is the composition of t and q (t is the map taking any element l of L to the pair $(r(l), s(l))$).

The category-theoretic approach *defines* a product for groups G and H to be any group K with homomorphisms p and q having such a universal property (invariably illustrated by a diagram when first introduced). In general, abstract category theory, this is how the notion of product is defined. It is not assumed that every category must have products, but many do, including not only the category of groups, but also that of topological spaces. (For instance, the product of a line and a circle is a cylinder.) And many results can be shown by a single proof to hold for all categories that do have products.

On this approach, the set-theoretic construction is "demoted," so to speak, from the definition of product to the demonstration that groups do have products in the category-theoretic sense, and its details may

in G, we must have $f(a) \cdot f(b) = f(c)$ in H. The contrast between forwards and backwards is Mac Lane's first example of a general contrast between *covariance* and *contravariance*, part of the beginning of an accumulation of a large vocabulary of new terminology connected with the new theory (though this particular pair had, in fact, already been in use elsewhere in mathematics in a not unrelated sense).

be forgotten after it has been used for this purpose, as so many other set-theoretic constructions are forgotten after serving their purpose, in the manner of the Zermelo and Von Neumann constructions of the natural numbers. Or rather, the category-theoretic perspective is that they *must* be forgotten, and any proof of further results about products be given solely on the basis of the category-theoretic universal-property definition. This approach "takes up slack" to a certain extent, in the sense that one does not need to settle on any convention about just how the product structure is to be created. We could have taken its elements to consist of pairs (h, g) rather than pairs (g, h), for example.

Here the category-theoretic approach seems entirely natural, in a non-technical sense of the term. In other cases, it is a bit less so. Even before set theory, even before Cantor, when groups were not yet thought of "in the abstract" but always as groups of permutations or transformations, one sometimes considered *all* relevant permutations and transformations and got a group, and sometimes considered only some, and got a subgroup. In set-theoretic terms, H is a subgroup of G if the set of elements of H is a subset of the set of elements of G, and the operation on H is just the operation of G restricted in application to that subset.

This traditional and perfectly natural notion has to be given up on a strictly category-theoretic approach. It is replaced by the notion of an *embedding f* of one group K in another group G, which in the pre-category-theoretic approach would be simply an isomorphism from K to a subgroup H of G. Instead of saying that a certain Galois group has four distinct (though necessarily isomorphic) noncyclic subgroups H_1, H_2, H_3, H_4 of order 4, one says that given a noncyclic group K of order 4, there are four distinct embeddings f_1, f_2, f_3, f_4 of it into G. Here the notion of "embedding" has to be defined as a morphism with certain additional (universal) properties definable solely in terms of composition of morphisms. For that matter, on a strictly category-theoretic approach, isomorphism has to be similarly defined, as well as "cyclic" and even "order four", since on a strictly category-theoretic approach one does not count the elements of a structure, or even mention that a structure *has* elements, but only counts morphisms.

There seems to be a considerable initial investment involved in redefining everything in category-theoretic terms, though in fact the definition of "embedding" (or "monomorphism") and "isomorphism" has to be given only once, for *all* categories. Only the notion of morphism, represented in diagrams by arrows, and of the "objects" from which the morphisms, from which the arrows begin and in which they end, need to be separately defined for each category. And a surprising number of facts can be proved at the category-theoretic level of abstraction, and applied to any branch of mathematics one has learned to view category-theoretically.

Moreover, the notion of category is nothing if not general. Mac Lane realized that, even when the objects of the category are structures, that is, sets with additional structure, the morphisms between them do not need to be functions with additional properties; and more importantly, the objects don't *have* to be structures. When category theory is developed in the abstract, nothing is said about what the objects or the arrows of the category are. Indeed, it is not really necessary to mention the objects at all: The whole story can be told in terms of the arrows.

A category is a structure consisting of arrows, some of which can be composed (heuristically, those where the object that is one arrow's target is the same as the object that is the other arrow's source), and for which composition obeys certain laws, set forth in the original paper of Eilenberg and Mac Lane. Just as in abstract group theory special kinds of groups are considered, such as the Abelian groups, so in abstract category theory special kinds of categories are considered, one of which kind is in fact constituted by Abelian categories.[2]

Given questions in, say, group theory, one can consider the category of groups, or all Abelian groups, or all groups of some other relevant kind, apply general theorems about all categories, or all categories of a certain kind taking in the category of immediate interest, to reach conclusions about that category, and hence ultimately conclusions about the groups of original interest. Those results in turn may be applied to groups of permutations or transformations to obtain results in some more "concrete" branch of mathematics. So schematically, now there are three levels of abstraction: the bottom level of algebraic permutations or geometric transformations, which as mathematicians rather than philosophers use the term is "concrete," the level of structures such as groups and topological spaces, and then the level of categories. This is not the only way

---

[2] These are not categories in which the composition operation obeys some analog of the Abelian or commutative law, but ones that share certain key features with the category of Abelian groups.

categories can be seen as coming up in different mathematical contexts, but it is the most basic.[3]

As for philosophical structuralism, I promised earlier, after observing that one kind of mathematical object evidently *not* treated "structuralistically" in mathematical practice was precisely *structures*, such as groups and rings and fields, to revisit the issue when there would be something more positive to say; and now there is. For *any* mathematical object, there is a level of abstraction at which it does appear to be just a featureless point in a structure, and there is a level of abstraction at which it disappears altogether. For a permutation of the roots of an equation, at the concrete level where Galois theory is applied to construction problems or the solution of equations by radicals, it appears as an object with a definite nature, namely, that of a permutation, a function carrying inputs to outputs subject to certain conditions. There is a higher level of abstraction, that of group theory in the abstract, where it and other group elements appear only as featureless points in structures, their groups (or to put it less mystically, a level where they are mentioned without mentioning any features, but only their relations to other elements in a group). At this level, the groups themselves still appear as objects with a definite nature, as do homomorphisms

---

[3] Let me say just a word or two about contexts in which categories arise more directly. Not all the "structures" considered in contemporary mathematics are just sets with operations, like groups and rings and fields, or sets with distinguished families of subsets, like topological spaces. In particular, many more complications are involved already with the notion of *manifold* found in algebraic topology and differential topology and differential geometry. A manifold is a topological space with a lot of additional apparatus reflecting the requirement of being "locally Euclidean." Around every point there must be a little open set or "neighborhood" in which the space looks almost like Euclidean n-dimensional space for some n, in particular in the sense that it must be possible to assign *coordinates* to points in the neighborhood. But one must also indicate how changes of coordinate take place between different possible coordinate systems on the same open set, and between distinct but overlapping open sets. So already one has a set-up with something fairly complicated associated to open sets, and mappings connecting these complicated extras. Categories of "objects" or "structures" connected by "arrows" or "morphisms" are clearly lurking nearby. They really come into their own in algebraic geometry's higher reaches, with Grothendieck's notion of a *scheme*, but since it takes a year or two of graduate study before one even gets to the definition of *scheme*, I will make no attempt to say more about it, except that there are lots of opportunities to bring in category theory once one reaches that stratospheric level, and in the SGA (Séminaire de Géométrie Algébrique: Algebraic Geometry Seminar) series Grothendieck takes full advantage of such opportunities. The notion of *topos*, which is made prominent by SGA 4 (Artin et al. 1972), and whose name one cannot read very far in the categories-as-foundations literature without encountering, originated at this level. For our purposes, it will be enough to know, when we come to them, that a topos is, like an "Abelian category," a structure satisfying the Eilenberg–Mac Lane defining axioms for a category, plus some extra assumptions.

between groups, and similarly for rings and fields and so forth, namely their nature as structures with sets of elements and operations, or as operation-preserving functions between such sets of elements.

But there is a higher level of abstraction, that of the category of groups, and of rings and fields and so forth, which appear only as featureless points in structures, their categories (though this could also be put less mystically). At that level, the elements of the groups or rings or fields or whatever disappear entirely (or to put it less mystically, are never mentioned at all), while the categories still appear as objects with natures, namely, as systems of objects and arrows. There is a still higher level of abstraction, that of a category of all categories, but we are not quite ready for that. At any rate, each mathematical object does *sometimes* appear just as mystical structuralists conceive it, as a featureless point, and each mathematical object does *sometimes* appear just as hardheaded nominalists conceive it, as not there at all. *Only one cannot always remain at these high levels of abstraction, lest applicability be lost.*

Bourbaki Revisited

How widely useful the new, higher level of abstraction may be is a question that need not be entered into at this point, except to say that some members of the second generation of the Bourbaki group—it was a rule that each member had to withdraw at age 50, to be replaced by sporadic recruitment of younger mathematicians—found the approach useful in their own work, and achieved, partly by category-theoretic methods, results about non-category-theoretic problems that forced the mathematical community as a whole to sit up and take notice.[4]

By contrast, the vague ideas about "mother structures" floated in Bourbaki's architectural remarks have had hardly any visible influence even within the later Bourbaki volumes. Now the chronological sequence of events is that the first Bourbaki volume appeared in 1939, the Eilenberg–Mac Lane paper in 1945, and the heavily category-theoretic work of Grothendieck in the 1960s. So we cannot ask whether Bourbaki should have used categories from the beginning. It is only in retrospect that one can ask: Should categories replace structures as the culmination of the first volume of the

---

[4] I am thinking above all of the romantic figure of Alexandre Grothendieck, a paradigmatically difficult genius, who after making enormous contributions, eventually split from the Bourbaki group and ultimately renounced mathematics.

Bourbaki series? It seems to be a fairly widely held view that the answer must be affirmative, *so long as the Bourbaki principle is still adhered to, of always doing everything the first time around in the maximal degree of generality that will subsequently be needed.* But for many, this conclusion, that categories should have been there from the end of the first volume on, is a *reductio ad absurdum* of the Bourbaki maximal generality principle.

Certainly, it seems so from a pedagogical point of view; and though the Bourbaki volumes have in practice served more as reference works than as textbooks, pedagogy had been on the minds of the founders. To introduce categories as early as the first volume would seem unwise, and would be contrary to the pedagogical principles *of the main creator of category theory himself.* We can know this because Mac Lane is author or co-author of three textbooks, and we can observe his practice. If we do so, we find that none of his books introduces category theory very early.

His *Categories for the Working Mathematician* (1971) appears in a series of *graduate* texts and mentions advanced topics so freely that it is clearly not meant for undergraduates. Mac Lane has also written two books jointly with Garrett Birkhoff. The shorter, Birkhoff–Mac Lane, *Survey of Modern Algebra* (1941), written as an undergraduate text, but sometimes used in summer math camps for high-school students (I myself first learned abstract algebra from it in such a context), does not introduce category-theoretic methods at all. The longer, Mac Lane–Birkhoff, *Algebra* (1988), does introduce category-theoretic material. The following snippets (from the preface to the first edition, reproduced in later editions) reveal the authors' perspective:

[W]e hold that the general and abstract ideas needed should grow naturally from concrete instances. With this in view, it is fortunate that we do not need to begin with the general notion of a category. The most basic category is the category whose objects are all sets and whose morphisms are all functions (from one set to another); hence we can start Chapter I with sets . . . All the other categories we need are quite "concrete" ones—each object *A* in the category is a set (with some structure), and each morphism from an object *A* to an object *B* in the category is a function (one which preserves the structure) on the set *A* to the set *B*. (Mac Lane and Birkhoff 1988, p. viii)

They go on to say (speaking of the first edition) that they introduce the notion of concrete category in chapter 2 and postpone the general notion until chapter 14. By the third edition of 1988, there has been a postponement to chapters 4 and 15, respectively. Still, this is more category-theoretic material than virtually any other undergraduate algebra text, though the fact that the book has gone through multiple editions and reprintings is a sign of a significant level of success.

Reality Checks

The philosopher, not belonging to the mathematical culture, will not have much idea how category theory is regarded there, and will be and should be suspicious of statements on this head by interested parties.[5] The matter of the relative reputations of different branches of another field is not in general much our business as philosophers, but I do think something needs to be said on the matter in the present context, if only because the reader who follows up the literature on the contentious issues I will be taking up, reading as one should contributors on both sides, is bound to come across some rather extreme opinions one way and the other. Googling quickly turns up, for instance, "category theory was the greatest creation of twentieth-century mathematics" versus "the study of category theory for its own sake [is] surely one of the most sterile of all intellectual pursuits." And one may well wonder, in trying to decide how seriously to take an author who comes out with such a formulation, how widely the opinions thus expressed may be shared.

In my own experience, the strongest claims seem to come mostly from the pro- rather than the anti-category side. For many category theorists are imbued with a kind of missionary zeal for their subject that one rarely finds among, say, analytic number theorists—or set theorists. In the case of William Lawvere, the prime mover in the category-theory-as-foundations movement, there is an especially intense ideological fervor. The effect of his association of category theory with dialectical materialism, in particular, has been described by fellow category theorists Jean-Pierre Marquis and Gonzalo Reyes as follows:[6]

Thus, many category theorists came to the conclusion that it was preferable to avoid the logical aspect altogether . . . In particular, using logical techniques to

[5] I am speaking of the status of pure, abstract category theory, not of results obtained by the application of category theory by Grothendieck or others; for those, the award of major international prizes is a sufficient indication of how they are regarded.
[6] The specific context of Marquis's and Reyes's remarks is the approach of certain disciples of Lawvere, who they suggest may not have fully understood their master's remarks, to "logical" versus "geometrical" aspects of topos theory; for Marquis and Reyes themselves these are merely two facets, neither to be despised, of a multifaceted theory.

solve a problem was sometimes considered to be reactionary or fascist. It is even said that set theory was considered to be essentially bourgeois since it's founded on the relationship of belonging. (Marquis and Reyes 2012, 718)

This game of politicizing mathematics is one that anyone can learn to play, only most have better sense and better taste than to do so. Thus one could claim with equal justice (which is to say, with none) that the set-theoretic conception leaves the members of a group to form a genuine community of individuals collaborating and cooperating with each other, contributing each according to "his" abilities, and receiving each according to "his" needs; while, by contrast, the more abstract structuralist standpoint reduces them all to featureless, faceless units like proletarians on a capitalist assembly line, defined only by their places in the profit-generating industrial process, and while the still more alienating standpoint of category theory obliterates and exterminates them altogether, like prisoners in a *stalag*.

But setting the views of obvious ideologues aside, how does pure, abstract category theory compare in standing to, say, pure, abstract set theory? One statement that is safe to make is that it is possible to have a world-class mathematics department, in the strict sense of one consistently ranked among the top few in the world, without having among its dozens and dozens of faculty members any who is a full-time or half-time or even quarter-time specialist in either subject, pure set theory or pure category theory, leaving any and all teaching of those subjects entirely to associated faculty in other departments of the university. This characterization fits the mathematics department at my own university exactly.[7] Perhaps the intensity one sometimes finds in debate over set theory versus category is partly explained by Sayre's law: "In any dispute the intensity of feeling is inversely proportional to the value of the issues at stake. That is why academic politics are so bitter." Beside the claim about world-class departments, a couple of other statements seem to me almost equally safe to make.

Category theory arose in and around algebraic topology, and was soon employed in algebraic geometry, and to this day is more likely to be appreciated by those whose work is primarily in certain key branches

---

[7] I am the one who teaches set theory, and my colleague Hans Halvorson the one who teaches category theory.

of algebra or of geometry topology rather than, say, partial differential equations. That there should be this kind of variation in appreciation is what one would expect of a *tool*, which is always going to be more useful for some jobs than for others, as will be recognized by everyone except that proverbial man with a hammer to whom everything seems to be a nail.

Category theory is also more likely to be appreciated by mathematicians who are more in the theory-builder than the problem-solver mold, to use a distinction recently illuminatingly discussed by Fields medalist Timothy Gowers, who is quick to insist that it is a matter of priorities, not of exclusive interests (see Gowers 2000). And what is the difference? Well, consider Wiles's solution to the Fermat problem, which involved application of some very high theory. All mathematicians regard this as a very great achievement; but for some mathematicians what it does is to show, finally, that high theory was worth thinking about; while for other mathematicians what it does is to show, finally, that the Fermat problem was worth thinking about. The former group are the problem-solvers, the latter the theory-builders. That there should be more interest on the part of theory-builders than of problem-solvers in category theory is what one would expect of a tool that is highly abstract.

Beyond this, I would urge the reader, while in the mathematics library, to try the following exercise: Take a few recent issues of the *BAMS* and/ or some other general journal (not limited to a particular branch of mathematics), and leaf through the articles, looking on the one hand for set-theoretic notation and vocabulary, and on the other hand, for category-theoretic. I predict the reader will find that the language of set theory will be used in every single article, and the language of category theory in what amounts to a sizable minority: a larger minority than you would have found twenty years ago, but still a minority. This does not show you whether any higher results from the pure, abstract, higher theory of either subject, beyond what one would encounter in a first course, are being applied. I know enough about set theory to be able to say with confidence that for set theory this will almost certainly not be the case for more than perhaps one article in a year's run of a journal. I expect that it will be the case for a larger percentage of the articles that find category-theoretic language useful, but even so by no means for all, though here I cannot really speak from expert knowledge. The exercise should suffice to cure the reader of any tendency to think either that "abstract nonsense plays no role in real

mathematics" or that "the category-theoretic point of view pervades contemporary mathematics."[8]

One more thing the reader may be sure of: In any case where there
exist both a category-theoretic and a non-category-theoretic proof of the
same result, some are going to prefer the former, and others the latter.[9]
What one can generally say about either proof in such a case is what is said
in the old joke: "People who like this sort of thing will find this the sort of
thing they like."

A Common Confusion

There are enough real contrasts and potential conflicts between the
category-theoretic and set-theoretic perspectives that we do not need to con-
fuse the situation by bringing in conflicts that are merely apparent. A case in
point: Commentators sometimes write as if set theorists wished to impose
some kind of strait-jacket on category theory, preventing the natural formulation of its theses as pertaining to *all* structures of a given kind: *all* groups, *all*

---

[8]   Here "abstract nonsense" is a term attributed to Norman Steenrod, one of the immediate
precursors of category theory, for the kind of "diagram-chasing" and other arguments from
highly general considerations found in the background material out of which the work of
Eilenberg and Mac Lane developed. The term was later extended to category theory proper.
It was originally an affectionate nickname, though it is not always used as such today. Some
use "general abstract nonsense" or "generalized abstract nonsense" as synonyms; others take
these to refer specifically to topos theory or higher category theory.

[9]   Early in my career, shortly after I had, as a post-doc, written the chapter on Cohen's
method of "Forcing" in the *Handbook of Mathematical Logic* (Burgess 1977), I overheard
some conversation among category theorists about the work of Peter Freyd, who had
translated independence proofs by forcing into category-theoretic language. From a conventional point of view, the value of this work lies in extending the results to a wider class
of models, topoi. But what I heard people saying was, "Now this theorem has really been
proved," as if the proof for which Cohen received the Fields medal were not proof enough.
This is the sort of language one would expect to hear if, say, a computer-independent, "conceptual" proof were found for the four-color theorem; but I was shocked, and I suppose
the shock has colored my perception of category theory and category theorists to this day.
Goethe is quoted as saying of mathematicians that whenever you say something to them,
they translate it into their own language, and immediately it becomes something else. Other
mathematicians might say the same sort of thing about category theorists. I would define
the distinction between a *fan* of category theory and a *fanatic* as follows. A fan of category
theory is someone likely to believe that translating a problem into category-theoretic terms
may well, by revealing analogies with other areas of mathematics and in other ways, do
something to facilitate its solution. A fanatic is someone who counts the mere translation of
a theorem or a proof into category-theoretic terms as already in itself constituting some kind
of major contribution.

topological spaces, and for that matter, *all* sets, and *all* categories. But what in fact imposes limits here is the desire for consistency.

In set theory, one cannot consistently allow *both* an absolutely universal set of absolutely all sets, and absolute freedom to separate out from a given set the subset consisting of those of its elements satisfying some condition. Frege allowed absolute universality and absolute freedom to separate out subcollections in his theory of classes, and the Russell paradox was the result. It arises if one separates out from the class of absolutely all classes those satisfying the condition "*x* is not a member of itself"; this is the Russell class which must be a member of itself if and only if it is not. Orthodox set theory, embodied in ZFC, opts to retain absolute freedom to specify subsets, a decision reflected in Zermelo's axiom of separation: From any set u, for any condition ϕ, one can form the set of all x in u that satisfy ϕ, in symbols, $\{x \in u : \phi(x)\}$. Orthodox set theory therefore has to renounce the idea of a universal set. The opposite option has to some extent been explored, especially in Quine's set theory NF, but it seems to be considerably less convenient: There is a constant need for bookkeeping to check that the conditions one is working with are ones that can legitimately be used to separate a subset from a set.[10]

Now when expressing "architectural" ideas it does come naturally to speak of a category of all groups or all topological spaces. There are even variants of ZFC, the best known being called NBG (Von Neumann-Bernays-Gödel) and MK (Morse-Kelly), that allow a layer of "classes" too large to be sets on top of the universe of sets, and in this way to a degree permit thinking of all groups or all topological spaces as a single

---

[10] NF (from "New Foundations", the title of the 1937 paper of Quine in which it appeared) is by far the best-known version of the opposite to the orthodox approach. Its consistency is to this day considered rather problematic (at the time of this writing it is being reported, not for the first time, that a purported proof has been obtained but is still being vetted by specialists). An extension, proposed by Quine in the 1st edn of his *Mathematical Logic*, was found to be inconsistent; a restricted version in the 2nd edn, called ML, is known to be consistent if NF is; the story is told in the 2nd edn, Quine 1951. Another option, suggested by some philosophers of logic, is to adopt a "paraconsistent" logic, one that can somehow tolerate contradictions. I will state flatly that this option is unworkable, because once one moves beyond classical logic and its intuitionistic rival, one may say of any of the further logics on the market what Solomon Feferman famously said of one of them, that "nothing like sustained ordinary reasoning can be carried on" using them. This can be seen from the fact that, in contrast to classical and intuitionistic logicians, when advocates of such logics want to prove metatheorems about their own logics, they do not conform to those logics in their proofs, but make use of classical, or at least intuitionistic, modes of argument that on their own official view are generally invalid.

object. But this only postpones problems. For if one wants to start performing the usual sorts of operations category theorists perform on categories (for instance, looking at so-called *functor categories*), one is soon going to need more than classes, and if one admits "superclasses," more than those, too. Perhaps one doesn't need to perform such operations when simply musing about architecture, but they do tend to turn up in applications.

For applications, however, one doesn't need a category of literally *all* groups to begin with. It is always enough to have a category of "enough" groups, though how many is enough may vary from application to application, and a threat may loom that one will be required to do some bookkeeping to keep track of how many one is using in any given application. Grothendieck in SGA 4 found a way out of the difficulty. Simply assume one has a set of groups closed under all the operations one could conceivably need, and consider the category of groups in that set. More precisely, Grothendieck posits a "local universe" closed under all the operations provided for by the axioms of set theory (forming pairs, unions, power sets, and more) and considers the category of all groups in that universe. That category is too large to itself belong to the local universe, and standard category-theoretic constructions may lead to still larger categories. In order to be sure that any general theorems one may have proved about categories using the assumption of the existence of a local universe will apply to these large categories, one needs to assume that they, too, belong to a (larger) local universe. In fact, Grothendieck's hypothesis is that every set, however large, belongs to some local universe: The "global universe" of absolutely all sets is simply the union of increasingly large local universes.

Now there can be no question of proving this hypothesis, or even of proving the existence of a single local universe, from ZFC, and to this extent Grothendieck's proposal has the advantages of theft over honest toil. The reason that there cannot be a proof from ZFC (supposing ZFC to be consistent) is that a local universe would be a model of ZFC, and the existence of a model would imply the consistency of ZFC, and we know from Gödel that no theory (if consistent) can prove its own consistency. In fact, one can say a bit more.

Though ZFC cannot prove its own consistency, let alone the consistency of ZFC + CH, the result of adding the continuum hypothesis CH as a further postulate, it can prove the conditional or hypothetical or *relative* consistency statement that, *if* ZFC itself is consistent, adding CH will not create an inconsistency that wasn't there before. In fact, this much can be proved without using anything anywhere near the full resources of ZFC: It can be given a proof that an intuitionist and even a finitist could accept. Similar remarks apply to ZF and the axiom of choice AC. The relative consistency results of Gödel and Cohen tell us the following:

$$\text{Con ZFC} \to \text{Con ZFC} + \text{CH}$$
$$\text{Con ZFC} \to \text{Con ZFC} + \text{not-CH}$$
$$\text{Con ZF} \to \text{Con ZF} + \text{AC}$$
$$\text{Con ZF} \to \text{Con ZF} + \text{not-AC}$$

Indeed, one can have any combination one pleases of options for or against AC and CH, so that if the basic ZF is consistent, so will be ZF ± AC ± CH. All these theories have the same "consistency strength," as is said.

The same cannot be said for what we may call $ZFCG_1$, or ZFC plus the assumption that one Grothendieck local universe exists, let alone for ZFCG, or ZFC plus the full Grothendieck hypothesis that every set belongs to a local universe. $ZFCG_1$ implies Con ZFC. This fact has already been stated in slightly different terminology. If

$$\text{Con ZFC} \to \text{Con ZFCG}_1$$

were finitistically provable, or intuitionistically provable, or even just "classically" provable in the sense of being provable in ZFC, or indeed even just provable in $ZFCG_1$, then Con $ZFCG_1$ would be provable in $ZFCG_1$, and by Gödel's theorem, $ZFCG_1$ would be inconsistent. $ZFCG_1$ is of "greater consistency strength" than ZFC, which is a euphemistic way of saying that it is inherently riskier. The full ZFCG is even more so.

Nonetheless, it cannot be said that set theorists on these grounds are forbidding category theorists to assume Grothendieck's hypothesis. (As if set theorists had such power to forbid or permit!) Quite the contrary, *set theorists have been inviting mathematicians to make this assumption ever since 1930, when Zermelo first formulated it.* In set theorists' jargon G_1 is the hypothesis that there exists an "inaccessible cardinal," and the full G is the hypothesis that there exist "cofinally many" such cardinals (that is, for every cardinal, there is a larger cardinal that is inaccessible). I know of no set theorist who believes that there is a heuristic picture motivating the axioms of ZFC who does not believe that the same picture motivates ZFCG. In particular, Gödel himself was firmly convinced of what non-set theorists know as "Grothendieck's hypothesis."

In any case, the opinion of the handful of experts who know both enough set theory and enough category theory to have an opinion seems to be that, whether or not Grothendieck was aware of the fact, which is as may be, for Grothendieck's purposes, Grothendieck's hypothesis is a convenience that it would be in principle a possible option, if not in practice an attractive one, to dispense with.

The reason is that there is a technical result in set theory, called the *reflection principle*, according to which, to put it roughly, while ZFC can't prove the existence of a set closed under *all* the operations it admits, it can prove the existence of one closed under *any finite number desired* of such operations. In making an application, if one knew in advance that one was going to need only the first 57 operations, say, then instead of assuming there is a local universe U it could be assumed that there is a local "almost-universe to the 57th degree" U_{57}.

Of course, in general one does *not* know in advance, so one would have to say, "Let U be an almost universe to the nth degree, where n remains to be determined," and then go back after one finishes to do the bookkeeping needed to determine the value of n needed, or to tell readers that the determination of the value of n is being left as an exercise for them.

An extended and sometimes polemic discussion on the online subscription list "FOM" (for "Foundations of Mathematics"), involving Harvey Friedman, Colin McLarty, and others, on whether Grothendieck's work involves "necessary uses of large cardinals" (inaccessibles being the smallest "large cardinals") appeared to converge towards the more or less irenic conclusion that the answer is probably "no" if by "necessary" one means absolutely logically unavoidable, but "yes" if "necessary" is taken in a less strict, more psychological sense. In Grothendieck's kind of work the intellectual faculties are being strained to their uttermost limit, and one doesn't want the distraction of any sort of bookkeeping requirements, any more than an acrobat walking a very high tightrope wants the distraction of gnats. There is a good deal more that could be said, but it is mainly of a technical nature: There has been an ongoing series of technical publications on the subject by McLarty (see especially 2010) and Feferman (such as his 1977 and 2013), though forthcoming work of Michael Ernst raises doubts about how far the latter approach can be pushed. Still in sum I think one can say that there is really no reason to see a significant conflict between set theory and category theory here—unless one is trying to pick a fight.[11]

Alpha and Omega

As I have already hinted, it seems to be the ambition of some category theorists that category theory should be *both* first *and* last in mathematics, or both bottom and top: Not content with recognition as the highest level of abstraction, they seek recognition as the deepest level of foundations as well, giving the mansion of mathematics an Escher-like structure, in which the attic is also the cellar. It is not, however, always easy to make out what is meant by "foundations" when the word is used by category theorists.

Now no one can claim a monopoly on the word "foundations." In this chapter, I have used it to refer to issues about what the primitives and postulates should be in a rigorous development of mathematics (but without any association with the kind of "foundationalism" that demands that the meaning and truth of the primitives and postulates should be "evident," since in the light of Gödel's work we know that trade-offs are inevitable between evidence and power). I claim no more than that my usage represents *one* well-established sense of the term.

There is another very well established usage that I have not yet mentioned, the use of "Foundations" as a label in American Mathematical

---

[11] Before leaving the topic on this more or less happy note, I should acknowledge that something I said earlier about provability in ZFC being the standard for mathematics journals may require a qualification. For the famous volume, SGA 4, in which Grothendieck advances his hypothesis, is widely cited, and many of the works citing it are widely cited in turn, and on the face of it Grothendieck is making an assumption going beyond ZFC; but an annotation to that effect is not being passed along down the chain of citations. That means that there may be papers in the *Annals of Mathematics* and elsewhere that, if one just went by the chain of their citations, would be "infected" with large cardinals, if only very small ones that set theorists feel quite confident about, and if only in a way that according to experts could in principle be eliminated. All this is to say that, if the *Annals* really has a semi-official policy of accepting only proofs not requiring resources beyond ZFC, then there may have been breaches of policy de jure, though not de facto. But the matter is simply too technical to be gone into in detail, or even adequately summarized, in a work of the present kind. Also it should be noted that nothing "irenic" said in the text is meant to deny that occasionally later category theorists have gotten themselves into set-theoretic difficulties, or proposed category-theoretic constructions that would require larger "large cardinals" than inaccessibles. Examples are provided by Mathias in another anti-Bourbaki position piece (see Mathias 2014), as well as in technical publications.

Society (AMS) subject classifications. On this usage, "foundations" covers not only traditional questions about the first principles for a rigorous codification of mathematics, but also *any mathematical developments whatsoever making significant use of any of the methods originally developed in connection with such traditional questions, regardless of the nature of intended applications, if any.* This expanded usage exactly parallels the expanded usage of "geometry" beyond traditional concern with the structure of the space around us, in which we live and move, to cover *any mathematical developments whatsoever making significant use of any of the methods originally developed in connection with such traditional questions, regardless of the nature of intended applications if any.* In the AMS sense, all the work of category theorists that I will be discussing is unambiguously work in "foundations"; but much work that is "foundational" in the AMS sense has virtually nothing to do with "foundational" issues in the sense with which I have been concerned in this essay.

By contrast, if the prediction that mathematical journals in the future will require documentation by formal proofs of the sort produced by computerized proof-assistants comes true, then a question about which implementation of which format is to be required will open up. Since the different implementations of the different formats differ in their starting points, their formal languages and formal axioms, this would be a foundational question in the more or less traditional sense in which I have been using the expression.[12]

---

[12] The potential for confusion lurking in the ambiguity of "foundations" for a traditional sense and the AMS classification sense may be involved in some discussions of Vladimir Voevodsky's "univalent foundations," in which there has recently been an explosion of interest, with a special year devoted to the topic at the Institute for Advanced Study, followed by a heavy investment by the US Department of Defense (!) in further development at Carnegie-Mellon University under Steve Awodey. Here there is in the background a variant of something definitely foundational in a traditional sense, Martin-Löf constructive type theory, long the most actively pursued constructive alternative to classical mathematics. An assumption not traditionally accepted by constructivists, called the "univalence axiom," is added, the mechanical proof-assistant Coq is deployed, and above all an ingenious reinterpretation of the formalism is proposed. The reinterpretation in effect turns deductions in type theory into calculations in what is known as homotopy theory (to which Voevodsky has famously contributed in the past), with category-theoretic ideas being brought in at this part of the story. On the face of it, this looks like harnessing a formalism introduced to express constructivist ideas to do technical work in classical mathematics. We are certainly still in the "foundational" realm in the AMS classification sense, but are we in the foundational realm in any more traditional sense? Well, some commentators are already speaking of "univalent foundations" as if they were the same kind of thing as "set-theoretic foundations." Some even speak of the former as a rival destined to supplant the latter, and this within a very few years.

The present question is, to what extent do category theorists who speak of "foundations" have in mind something like "foundations" as understood in this chapter? To what extent are they speaking of the axioms of some kind of theory of categories replacing the standard axioms of set theory as the starting point for rigorous deduction of the rest of mathematics? The only thing that is really clear is that not all category theorists mean the same thing by "foundations," and hence not all category theorists mean by "foundations" what I have meant by it in this chapter.

That different category theorists mean different things by "foundations" is clear just from the abstracts of the various contributions to the recent (May 2013) University of California at Irvine Category-Theoretic Foundations of Mathematics Workshop, organized by Michael Ernst.[13] Indeed, we have here another demonstration of just how diverse a range of topics, each with an undeniable interest of its own, can be fitted under what naively one would suppose was a pretty specific title for an academic conference, at least in philosophy and related areas. It would be an enormous as well as a thankless task to attempt to sort out all the different things that are meant by "foundations" by the various participants in this conference.

That different category theorists mean different things by "foundations" is also clear, and at a level more accessible to non-specialists, from an exchange in 2003–5 in the pages of *Philosophia Mathematica*, a journal whose publisher, Oxford University Press, advertises it as the only journal in the world devoted exclusively to philosophy of mathematics. The participants in the exchange were three members of the journal's large editorial board. (The present author is another.) Hellman, the modal structuralist, opened with a critical piece 2003 on category theory as a foundation for mathematics, and specifically as a framework for mathematical structuralism, which drew responses from the category theorists Awodey 2004 and McLarty 2004 and 2005.

This claim is perhaps best understood as a prediction that in the future formalized proofs in the type-theoretic framework with the univalence assumption is what mathematical journals will be requiring. As to that, time will tell. The very mixed reactions by prominent mathematicians to the quite modest proposals of Jaffe and Quinn about tightening up rigor suggest that Homotopy Type Theory as "foundation" rather than mere tool may be a hard sell in the mathematical community at large.

[13]   At the time of writing, the conference program, with abstracts of talks, is still available online, <http://www.lps.uci.edu/node/15355>.

Except for some recurrence of terminology between their two articles, one would hardly guess that Awodey and McLarty are speaking of the same subject. They certainly cannot both mean the same thing by "foundations" and so at most one could mean something relevant to "foundations" as I have been discussing it in this chapter. (And one does, and I will be discussing his views later in this chapter.)

Lawvere, an enormously influential figure among category theorists, and especially among those with "foundational" ambitions in one sense or another, *sometimes* uses "foundations" in something like the sense that has concerned me. To be sure, even when he does, he does not *always* attach the kind of special importance that Bourbaki would attach to having a single, unified framework. He therefore proposes not one but several "foundational" systems, two at least of which have been extensively discussed: ETCS, the "elementary theory of the category of sets," and CCAF "the category of all categories as a foundation" (see Lawvere 2005, 1966).

At least a few other important category theorists have adopted a standpoint even closer to that adopted here, in effect nominating a single theory (usually some version or variant of one of the two just mentioned) as a substitute for ZFC. The most prominent examples are perhaps one of the chief founders of category theory, Mac Lane, and one of the two *Philosophia Mathematica* editors just mentioned, McLarty.[14] It is mainly with these two that I am concerned here. In particular, Mac Lane and McLarty are both concerned to advocate some version or variant of ETCS.

Let me, however, first take up CCAF, the more exciting or at any rate the more purely category-theoretic approach. Here, all current proposals modify in one respect or another Lawvere's pioneering approach, about which doubts were first raised in a review by John R. Isbell 1967. McLarty (see his commentary in Lawvere 2005) speaks of later workers pursuing Lawvere's ideas, but others might speak of clarifying Lawvere's obscurities or correcting Lawvere's errors. Blanc and Preller 1975, for instance begin their cited paper with the words, "It is long known that Lawvere's theory ... does not work," and go on to describe some of his "theorems" as "non-theorems."

The fundamental difficulty is as follows. Switching from a set-theoretic to a category-theoretic perspective in no way removes the inconsistency

---

[14]   See Mac Lane and Birkhoff 1986 and McLarty 2007 and 2010, in addition to other works already cited, as well as Mathias 2001.

between absolute generality and absolute freedom to specify subcases. An absolutely general category of absolutely all categories would be a category that is among its own objects. There may be other categories of this kind, but most categories are not like this. The category of groups is no group, for instance. Absolute freedom to specify subcases would allow us to define a subcategory of the category of all categories consisting of those categories that are *not* among their own elements. And then we would face an exact replica of the Russell paradox. It seems that only a certain lack of clarity about just what was being proposed prevents us from saying that Lawvere has proposed a clearly inconsistent theory.

In the case of set theory, the orthodox line represented by ZFC retains absolute freedom to specify subcases and gives up absolute universality; the heterodox line represented by Quine's NF among other alternatives goes in the other direction and remains, as I have said, of doubtful status. The theories of a category of categories that have been developed in the wake of Lawvere's original work generally tend to follow the ZFC rather than the NF route.[15]

It would be misleading to think, however, that it is solely or even mainly logical worries about consistency that have placed the development of a "foundational" theory of a category of categories on the back burner. Strictly category-theoretic considerations have also played a role (beginning with the observation that it is not easy to define the original motivating notion of "natural transformation" purely in terms of a set-up with categories as the objects and functors the morphisms). Though the most relevant developments within category theory are ferociously technical, I will have to attempt to say something about them.

As was mentioned in connection with the category-theoretic approach to products, category theory "takes up some slack" in set-theoretic approaches, which is to say, it eliminates the need to make certain arbitrary choices (for instance, of a definition of ordered pair). Many category theorists very much pride themselves on this aspect of category theory, and cite it as a respect in which category theory is superior to set theory. The more candid concede, however, that there may be a little slack left even in standard category theory. I cannot here undertake a serious introduction to higher category theory, and so can only gesture

---

[15] In particular, this is true of efforts in McLarty 1991 in this direction, which are comparatively modest in aim.

in the direction of the problem, by saying that it lies in the apparent fact that, though the general notion of isomorphism of structures applies to categories as much as to groups or anything else, isomorphism is not really the "right" notion of equivalence for categories, the way it is for groups.

This consideration is one that has helped to motivate the introduction of what are called 2-categories, with categories in the original sense now being called 1-categories. The new 2-categories are considerably fancier objects than the already rather fancy old 1-categories, the axioms for 2-categories being considerably more complicated than the "Eilenberg–Mac Lane axioms" for 1-categories, which are already so complicated that I have not wanted to burden the reader with a formal statement of them. A 2-category involves additional apparatus for which CCAF and the better known among its variants make no direct provision. Nor would a theory C2CAF be enough to satisfy all, since the same kind of considerations that motivated introducing 2-categories seem to call for 3-categories, and more generally n-categories.

We are getting here into the territory of some very complicated apparatus. The quip among the unsympathetic runs "n-categories are all very fine; the only trouble is that in order to understand them one must first have understood $(n + 1)$-categories." The problem would perhaps be solved if one could formulate a suitable all-encompassing notion of ∞-category, with no slack left in it at all, along with a theory C∞CAF; but consensus has, I take it, not yet been achieved as to how this should be done.

Indeed, there is "a little cloud . . . like a man's hand" visible on the horizon: There is reason to fear that it may not be possible, following the lines currently being pursued, to arrive at more than a notion of ω-category, with still some slack left, inviting the introduction of $(\omega + 1)$-categories, $(\omega + 2)$-categories, and more. In that case, category theory would seem to stand in need of help from some auxiliary theory studying the nature and extent of the sequence

$$1, 2, 3, \ldots \omega, \omega + 1, \omega + 2, \ldots$$

And of course, there does actually exist such a theory: It is called *set theory*.

Turning now to ETCS, before examining its pros and cons, let me note the most distinctive feature of the category-theoretic approach to set theory. Axiomatic set theory is concerned with sets as Cantor conceived them, unities made out of pluralities, their elements. ETCS is, as its full name indicates, concerned with the *category* of sets. Where in the category of groups we have groups and homomorphisms, and in the category of topological spaces we have topological spaces and continuous functions, so with the category of sets we have sets and functions. The example is always mentioned in expositions of category theory, because it gives the simplest examples of functors: the "forgetful functors" from the category of groups and from the category of topological spaces to the category of sets. This simply takes each set-plus-multiplication-operation or set-plus-family-of-open-subsets to the underlying set.

Now we have already seen that the elements of a group and the points of a space are in effect "not there" from a category-theoretic point of view, and the same is true for the elements of a set. Category theory does away with them, but not without offering something in their place. Given a unit set u, the elements of any other set x correspond to functions from u to x, and on a category-theoretic approach those functions are the shadows or ghosts of the departed elements.

The prospect of developing some sort of translation back and forth between a Cantorian–Zermelodic, element-based theory of sets, and a categorical, function-based theory depends in the one direction on these shadows or ghosts, and in the other direction on the admittedly rather artificial identification of functions with sets of ordered pairs and of ordered pairs with certain sets.

As I have briefly mentioned, once one has the homomorphisms and the continuous functions, from a category-theoretic point of view the groups and topological spaces are in principle superfluous. Likewise with the category of sets: It is the functions that matter, and not the sets that the functions go "from" and "to." The axiomatization of set theory on the basis of functions rather than sets was actually experimented with by Von Neumann, back in the days when the optimal axiomatization of set theory was still being sought, so the proposal of ETCS to axiomatize the universe of functions and composition of functions, rather than the universe of sets and elementhood, is not entirely unprecedented.

Yet there have been some fairly influential objections. McLarty, who has probably thought longer and harder about these matters than anyone, has developed what seem to me pretty effective responses to most of these. Let me go through a half-dozen of them.

First Objection

ETCS is too weak. In fact, ETCS is mutually interpretable or intertranslatable with a weakened version of Zermelo's set theory with choice called BSZC (for "bounded-separation Zermelo set theory with choice").[16] This theory is known to be equivalent to Ramsey's simple theory of types. So acceptance of the proposal to base mathematics on ETCS would involve jettisoning all mathematics that needs the extra power of ZFC over BSZC.

It might seem objectionable to advocate such a policy without a detailed survey of how much mathematics would be thrown overboard, and without much pondering of how much bookkeeping would be required to stay within bounds going forward. For some, what would be most objectionable here would be the uncollegiality of dismissing certain branches of mathematics as peripheral and hence dispensable rather than core branches, and perhaps even some results in core branches as peripheral and hence dispensable rather than core results. For others, what would be objectionable here would be the irresponsibility of simply assuming that there will be no costs to areas one considers core, without any serious prior analysis of the situation. Nor are these the only objectionable features.[17]

Now there may indeed be no concrete examples of uncontroversially indispensable core results in uncontroversially indispensable core areas for which the extra strength of ZFC over BSZC is indispensably logically necessary.[18] However, mathematicians like to state results in abstract, general form, going beyond the immediate needs of any foreseen concrete applications, and certainly at present they do so without stopping to check whether

---

[16] In this theory, the existence of $\{x \in u: \phi(x)\}$ is assumed not for all conditions ϕ, but only for logically fairly simple ones (involving only "bounded quantifiers").

[17] Kanamori 2012 argues that replacement, an axiom that would have to be given up, is a principle very much in a categorical or at least a structuralist spirit, a principle about the particular identity of mathematical objects not mattering, because they could always be replaced by others. He points to the result of Mathias that replacement is equivalent to the assumption that, *whatever* arbitrary choice one makes of how to define ordered pairs, there exists for all sets A and B the "Cartesian product" thereof, the set of all pairs (a, b) with a in A and b in B.

[18] Harvey Friedman has found fairly concrete results about so-called *Borel* sets, which play a role in real analysis, measure theory, and probability when done rigorously, and others in finite combinatorics. But these are not in general solutions to problems non-logicians working in those areas had actually posed prior to Friedman's work, and for the really skeptical core mathematician, that is grounds enough for suspicion, however disappointing such a reaction is to logicians.

they have used replacement in addition to separation, or have used some logically too-complex form of separation. Even if it were not too oner-ous a burden to have to keep track of such things in future, a good deal of work would be involved just in going over already accepted mathematics to purge it or provide it with modified proofs avoiding the appearance of relying on set-theoretic assumptions now deemed too strong. For noth-ing comes more naturally when working at an abstract, general level than to define inductively or recursively some mathematical object A_n for each natural number n, and then form the set of all such A_n, with a view to taking some kind of union or limit. And this step formally involves the axiom of replacement.[19]

McLarty, who has, as has been mentioned, carried out an analysis of the set-theoretic needs of Grothendieck's work,[20] and surely knows what effort would be involved in carrying out such a check across the whole of present-day mathematics, offers the alternative of working not with ETCS but with a strengthening thereof, ETCS-R, mutually interpretable with full ZFC. (The "R" here is for "replacement.")

But there have been objections to this proposal as well.

Second Objection

ECTS-R is too artificial. The replacement assumption simply isn't natural, in a non-technical sense, especially not from the category-theoretic point of view.

One possible response here would be *tu quoque*: Replacement isn't natural from a set-theoretic point of view, either, it might be claimed. Indeed, though many set theorists would argue that there is a heuristic picture underlying the axioms of ZFC and extensions thereof taking in inaccessibles and some other large cardinals, other logicians have some-times denied there really is such a picture and argued that there is no single motivating idea that covers *all* the axioms even of ZFC. And the best-known expression of this point of view, Boolos 1971, singles out the

---

[19] The use of replacement can be avoided if a background set B can be specified such that all A_n will be subsets of this B. Sometimes the existence of a suitable B is obvious enough to be "left to the reader as an exercise," but sometimes more work is involved to find a B and free a proof from ostensible dependence on replacement. More work will be needed if there are restrictions on separation to be observed as well.

[20] It is at the time of writing still unpublished, but is at least available online, <http://arxiv.org/pdf/1102.1773v4.pdf>.

axiom of replacement as poorly motivated on the heuristic picture of sets most often cited.

This is a matter that cannot be gone into without going into technicalities, and there is really no need to go further at this stage, since McLarty's primary response to the unnaturalness objection is along quite other lines. Not to put too fine a point on it, the axioms of ETCS-R are an extension of the axioms defining a kind of category called a topos, which originally came to be studied in connection with Grothendieck-style abstract algebraic geometry. It may be that, from a purely geometric point of view, the additional axioms needed to get to ETCS and especially the further replacement axiom needed to get to ETCS-R, look artificial. But that is irrelevant if ETCS-R is being advocated, *not as a geometric theory, but as a set theory*, albeit a function-based one rather than an element-based one.

If replacement is natural on an element-based approach, as the objector presumably holds, then it is equally so on a function-based approach. So McLarty does not need a *tu quoque* response. This leaves another pair of objections, however, where a *tu quoque* response may be the most obvious, or only obvious, one available.

Third Objection

Even if ETCS or ETCS-R does not presuppose ZFC in a formal way, it presupposes on the pre-theoretic, motivational level a notion of set, or anyhow of collection, as a unity formed from a plurality, since the notion of category (of sets or of anything else) is always informally explained as the notion of a collection of (objects and) arrows.

McLarty or any category theorist would be on especially strong grounds in replying *tu quoque* to this objection, since *exactly* the same objection can be raised against ZFC, and in effect was raised by Skolem back in the 1920s. Skolem argued that axiomatic set theory presupposes a notion of "domain" in the sense of a collection of things the theory is about, the sets. To this set theorists have traditionally answered that, no, all that is presupposed is the things, the sets, and there seems to be absolutely no reason why a category theorist could not say likewise that ETCS or ETCS-R only presupposes some things, namely, functions, the arrows of the category of sets.

There remains, however, a not entirely unrelated objection.

Fourth Objection

It would be pedagogically disastrous to start the development of mathematics from category theory, since to this day students arrive at the notion of category only after considerable familiarity with the lower level of abstraction represented by group theory or topology, and arrive at that level of abstraction only after considerable familiarity with more concrete mathematics.

Hence the wise restraint of Mac Lane and Birkhoff in their algebra text. The *tu quoque* response to this objection is only too obvious: It would equally be disastrous to start students with axiomatic set theory. The experiment of the "new math" in the 1960s, which introduced a modicum of explicitly set-theoretic talk into the lower elementary grades, is hardly regarded as a resounding success, after all.

The real response to the objection, however, is that whatever we are talking about when we are talking about a "foundation" for the mathematics of professional pure mathematicians, we are not talking about a new curriculum for kindergarten or pre-school. It is true that a few basic set-theoretic ideas—absolutely without technical jargon—may be involved in, say, the Montessori approach, and that it is hard to imagine how category-theoretic ideas could play a role at that level; but that level of instruction simply isn't what's at issue.

But, again, there remains a not entirely unrelated objection.

Fifth Objection

The development of mathematics on the basis of ETCS-R would be wholly parasitic on the development on the basis of ZFC.

Here, McLarty protests that it has never been the proposal of advocates of ETCS such as Mac Lane, nor is it his proposal, to begin by interpreting orthodox set theory in category theory and then proceed as in a bourbachique development on the basis of ZFC. On the contrary, the introduction of N and even of R would be done in a more direct, category-theoretic way, before the elementhood idea is introduced as an advanced topic (the way the category idea is introduced as an advanced topic on the orthodox set-theoretic approach).

There remains, however, yet one more objection, which may already have occurred to the reader who has followed the main lines of our earlier discussion on indifference.

Sixth Objection

Why bother? What mathematicians want in the way of foundations is, I have already suggested, something that *they don't need to think about.* Why, then, reopen an issue that has already been settled in a tolerable way? Why should mathematicians at work on raising the edifice of mathematics higher have to go back down into the basement or subbasement and rearrange things? McLarty's proposal, not just to derive the axioms of ZFC from ETCS-R, digging a category-theoretic subbasement under the present basement, but to remodel several of the lower floors—not just to add a chapter 0 before chapter 1, but to rewrite chapters 1 through 10—makes the program for change even less attractive.

Without looking in the basement, which working mathematicians evidently don't want to do, one can tell this much about what it is necessary and desirable to have down there. First, it is necessary to have enough to make provision for the usual operations by which mathematicians build new structures out of old. What this means is that what is down there must be, if not ZFC itself, *something in which one can interpret ZFC (or something very close to ZFC).* Second, it is desirable to have down there something of whose consistency we have a reasonable assurance. Since we could not in the short term have even inductive assurance of the consistency with anything wholly novel, and in any case might want something more than merely inductive insurances, what this means is that what is down there should be, if not ZFC itself, *something that can be interpreted in ZFC (or something very close to ZFC).*

We can say, therefore, if we don't want to be spending much time working down in the cellar, what is left down there *might as well be* set theory, as embodied in ZFC. If ETCS-R is mutually interpretable with ZFC, then indeed we can equally say that it *might as well be* category theory as embodied in ETCS-R. But to repeat the objection, so what?

I have suggested that the phenomena of mathematical practice that have inspired structuralist philosophies since Benacerraf's time should be viewed, not in the way the structuralists view them, as manifestations of the peculiar ontological nature of mathematical objects, but rather as examples of a kind of indifference that is only to be expected as an accompaniment of full rigor; and now I am suggesting that this same kind of indifference is likely to render the "sets *vs* categories" debate pointless. This last suggestion, however, reflects the perspective of the working mathematician who wishes nothing more than to avoid foundational issues.

What about the perspective of workers in foundations as contrasted with "core" mathematics? Here, the most natural reaction to the "sets *vs* categories" debate seems to me to be "Why not both?" In other areas of mathematics, the availability of different formulations of a problem, say geometrical and algebraic, is considered an advantage, since it means two different kinds of tools would be at hand in seeking a solution. The equivalence of ETCS-R with ZFC, and of ETCS with BZST, suggests the possibility, in foundational work, of passing back and forth between different formulations, set-theoretic and category-theoretic. How could this not be an advantage?

Everyone should be happy to witness traditional set-theoretic approaches and alternative category-theoretic approaches competing to see which can contribute more to identifying questions independent of or undecided by ZFC/ETCS-R, and discovering implications among different answers to such questions, and proposing assumptions that in set-theoretic or category-theoretic formulation, as the case may be, might be added to ZFC or to ETCS-R, as the case may be, in order to settle some such questions. By all means let the two approaches compete to see who can contribute most to that kind of common intellectual endeavor. That is a very different, potentially very much more productive, kind of competition than competition for the rather hollow honor of being considered "the" foundation.

Before closing, I would like to say a few words about the activities going on down in the cellar, a cellar that in practice few category theorists and fewer core mathematicians ever visit. I would like to say something about what category theorists would have to involve themselves in if they sought to become involved in truly "foundational" work in the sense in which I have been using the term.

Foundations and Foundations* and Foundations†

Within the area of "foundations" as I understand it I would like to distinguish two from among the several kinds of activities now going on. I will use as labels for the two *vertical* foundations and *horizontal* foundations, or foundations* and foundations†. After describing these two areas of activity I will very briefly consider the possible bearing of structuralism—which may have been largely lost sight of during our extended discussion of category theory, but with which we are not quite completely done—on the more contentious and controversial of the two.

What I will call *vertical* foundations, or foundations*, is concerned with the trade-offs Gödel teaches us are inevitable between evidence and power. In more technical terms, it is concerned with elaborating a *scale of relative consistency strength*, and placing various "foundational" theories, or sets of primitives and postulates, on this scale.

For present purposes, we may say that theory A has *greater consistency strength* than theory B when A can prove the consistency of B, but not vice versa. Under fairly general circumstances this means that it is provable (from very weak background assumptions) that if A is consistent, then so is B, but not conversely; and under fairly general circumstances it also means that (with very weak background resources) B can be interpreted in A but not vice versa.

It is easy enough for those versed in Gödelian methods to contrive examples of theories of incomparable consistency strength, where neither of A or B proves the consistency of the other, and where neither implication

$$\text{Con } A \rightarrow \text{Con } B \qquad \text{Con } B \rightarrow \text{Con } A$$

is provable, and where neither theory is interpretable in the other. But the remarkable result of several decades of work by set theorists and proof theorists and other logicians on foundations* is that *there are no naturally occurring examples of this kind*. Innumerable "foundational" theories have been suggested, and with a very few exceptions all have been placed on a scale of systems that is *linearly ordered* by consistency strength.[21]

For instance, systems naturally embodying finitism, constructivism, and predicativism have been devised, and found to be of strictly increasing consistency strength, all far below that of orthodox set theory. Many formalisms proposed as clarifications of Brouwer's rather obscure principles have also been proposed and placed. And in the other direction,

---

[21] A summary account of the scale is given towards the end of ch. 1 in Burgess 2005. Szmielew-Tarski set theory, to be mentioned shortly, is treated in ch. 2.

a large number of stronger and stronger proposed additions to the ZFC axioms have been considered and placed on the scale in higher and higher positions.

The theories in the upper ranges of the scale are set theories. Those in the middle are often formulated as theories just of natural numbers and sets of natural numbers, but it seems they could all be reformulated as set theories weaker than Zermelo's or Ramsey's. The weaker theories on the scale have traditionally been formulated just as "arithmetics," theories of natural numbers, but at least in many key cases it is known that these can be reformulated as set theories, too, very weak ones, to facilitate comparison. As we have seen, some at least of the set theories have equivalent category-theoretic formulations; most likely other systems, further down on the scale, could be reformulated category-theoretically as well, but the matter does not seem to have been investigated in detail as yet.

The weakest system generally considered is usually presented in intermediate-level logic textbooks as "Robinson arithmetic," but it is equivalent to what is known as "Szmielew-Tarski set theory" (after Tarski and his student Wanda Szmielew). This set theory provides only for the existence of the empty set $\emptyset$, and for the existence for any set x and any y of the set $x \cup \{y\}$ consisting of x with the single additional element y adjoined. The natural numbers 0, 1, 2, can be introduced using either Zermelo's or Von Neumann's definition, but there is no infinite set of all natural numbers, and no induction. Nonetheless, a surprising amount of mathematics can be done within the system.

The stronger theories on the scale add to ZFC assumptions about "large cardinals," of which inaccessibles are the smallest. But a word of clarification will be in order about what is meant by a *large cardinal axiom*. The expression does not have a rigorous formal definition. Cantor showed that there is no largest cardinal, and a naive reading of the label "large cardinal" might suggest that any cardinal larger than a large cardinal is a large cardinal. Actually, the expression is so understood that to count as a large cardinal a cardinal must tower above all smaller cardinals in something like the way in which the transfinite, starting with $\aleph_0$, towers over the finite, or the absolutely infinite or too-big-to-form-a-set towers over the transfinite; and on this understanding there are many cardinals larger than any given large cardinal that are not enough larger to count as "large cardinals."

The strongest set theories commonly considered at present are what are known as systems of ZFC *plus rank-into-rank large cardinals*, but those who like to live dangerously are forever proposing new experiments. Dozens and dozens of intermediate systems have been identified. The Grothendieck hypothesis, which to many mathematicians may appear daring, is to hardened set theorists nothing more than ZFC + ε.

Inevitably, there are skeptics who don't believe the scale is all there. That is to say, there have been, and this for a half-century at least, occasional workers in foundations and occasional outsiders who have expressed the hunch that one or another system on the scale is actually inconsistent, so that all the theories above it are inconsistent as well, and the top part of the scale is not really present.

In view of Gödel's theorem on the unprovability of consistency there has been nothing much to say to such suggestions except what there has already been occasion to say in other contexts: Time will tell. Still, even skeptics who discount the claims of some set theorists that there is a clear enough heuristic picture behind the formal axioms—what the picture is we will come to shortly—must concede that there is at least considerable inductive evidence of consistency, in the form of experimentation with strong systems that has failed to lead to contradiction. And we have survived a couple of scares.

At least twice respected workers in foundations have not merely had suspicions of inconsistency, but have convinced themselves that they had *proofs* of inconsistency, somewhere along the scale. Moreover, they remained convinced for long enough that they have written up and (privately) circulated their arguments. Fortunately, when inspected by colleagues, the purported deductions of contradictions have proved to be fallacious.

There has been only one case of a strengthening of ZFC proposed by an orthodox set theorist, widely and sympathetically received for some time by other set theorists, that has turned out to be inconsistent; and it was from the beginning recognized as stronger than all other proposals and hence more dangerous. And besides, all this was a long time ago now, way back in the 1960s of glorious memory.

The behavior of set theorists in those heady times resembled that of the main character in one of the fairy tales of the Brothers Grimm, "The Fisherman and His Wife." In the story, a poor fisherman catches a flounder but releases it when it begs him to. On returning to the pigsty he shared with his wife, she points out that if the flounder could speak, it must have magic powers, and so he should have asked for some reward before he let it go. She sends him back to ask the flounder for better lodgings, he goes down to the sea and shouts out his request for a nice cottage, and returns to find his wife standing at the door of it. Not content for long, the wife soon has him going back, under ever stormier conditions, to ask the flounder to make his wife a noble, a king, the emperor, and the pope, and all the requests are granted. But when she asks to be given the god-like power to make the sun and moon rise and set, the flounder says, "Then you must go back to the pigsty," and so indeed they do. ZFC is not exactly a pigsty, but inaccessible cardinals are a nice cottage, what are known as weakly compact cardinals a noble's castle, measurable cardinals the king's or emperor's palace, Woodin cardinals the pope's cathedral—and the Reinhardt cardinal the power to move the sun and moon.

It was shot down by Kenneth Kunen in 1971. And yet, Kunen's proof made essential use of AC, and the inconsistency of ZF (without AC) plus a Reinhardt cardinal has not been demonstrated, suggesting there may be a way to extend the consistency-strength hierarchy beyond all theories of the form ZFC plus a large cardinal.

In contrast to work in foundations*, surveying the scale that runs from Robinson/Szmielew-Tarski to just short of ZFC plus a Reinhardt cardinal, and perhaps beyond, there is an area of activity that I will call *horizontal* foundations or foundations†, concerned with differences between theories at the same point on the scale of consistency strength. An example is the case of the continuum hypothesis CH, which is to say, the case of the choice between ZFC + CH and ZFC + not-CH.

Originally, foundations* and foundations† were not distinct enterprises. For programmatic remarks of Gödel 1947 sent set theorists off on a project of trying to settle questions undecided by ZFC using stronger and stronger large cardinal hypotheses, whose addition to ZFC is what produces the upper reaches of the consistency-strength hierarchy.

This program had one very notable success. Cantor was originally the creator of not one but two theories, a theory of sets of real numbers or linear points, and then a theory of arbitrary sets of arbitrary elements. Subsequent work has distinguished *three* rather than *two* levels: (i) *descriptive* set theory, concerned with special sets of real numbers, those obtainable by beginning with simple sets like intervals and then performing simple operations (taking complements, unions or intersections, images under continuous functions); such operations lead to the Borel sets already briefly mentioned, and then beyond to what are known as analytic and co-analytic and projective sets, investigated by the Polish and Russian schools of the 1920s and 1930s; (ii) problems such as CH that are about *arbitrary* sets of real numbers; (iii) problems belonging to what is called *combinatorial set theory* or *infinitary combinatorics*, which is largely concerned with sets still larger than the set of real numbers, and stands to the arithmetic of transfinite cardinal and ordinal numbers somewhat as finite combinatorics stands to the arithmetic of natural numbers.

A number of questions at the level of descriptive theory that the schools of the interwar years were unable to resolve came in the wake of the work of Gödel and Cohen on CH shown to have the same character as that hypothesis, which is to say, shown to be undecided by ZFC. This was in work of Levy, Solovay, and others in the immediate wake of Cohen. But it somewhat later transpired that assuming enough large cardinals—so-called *Woodin cardinals*—resolves these questions, and moreover has been shown to be, in a sense that can be made precise, *just enough* to resolve them. All this was shown by work of a "cabal" of California set theorists, culminating in the results of D. A. Martin and John Steel, and of Hugh Woodin (after whom the cardinals are named).

Gödel had hoped that large cardinals could similarly decide CH. The hope had been that, though ZFC + CH and ZFC + not-CH are "equiconsistent," for some suitable large cardinal assumption LC it would be found that one of ZFC + LC + CH and ZFC + LC + not-CH would be *inconsistent*. If the former, then ZFC + LC would settle CH in the negative, implying not-CH; if the latter, then ZFC + LC would settle CH in the affirmative, proving it. (Gödel's expectation was for a negative resolution, and at least for a time, his expectation was more specifically that one would be able to prove $c = \aleph_2$.) Unfortunately, the early applications of Cohen's methods showed that CH is left undecided by ZFC + LC, essentially regardless of how strong a large cardinal hypothesis LC one considers.

At present, there is a division between two opinions. On one side stand those like Woodin (whom I mentioned as a part of the California cabal, though he has since left the West for the East Coast), who continue to work in foundations† and seek a decision on CH. On the other side stand *defeatists,* who take CH to be permanently or absolutely undecidable. This division often is (though I will argue that it need not be) cast as a division between those who believe that there is a single "intended interpretation" of the axioms of set theory, with respect to which a statement like CH is true or false, and those who believe that all there are in set theory are *multiple* models of the axioms, variously related to each other. The latter view is sometimes seen as a set-theoretic analogue of the view of some speculative cosmologists that we live in a physical *multiverse* and some speculative metaphysicians that we live in a modal *multiverse* of "possible worlds." Evidently, in contrast to work in foundations*, which can be viewed as a collection of purely technical results about relative consistency that are meaningful even from a finitist or stricter standpoint, work in foundations† has a philosophical aspect, or has philosophical presuppositions.

In particular, it is sometimes suggested that a structuralist philosophy of mathematics strongly favors a defeatist position. At first glance, this suggestion seems quite plausible, since after all structuralism does oppose the notion of a single "intended interpretation" in any area to which it is applied, while all attempts to go beyond a formal system of axioms like ZFC do seem to depend on appeal to an "intended interpretation" of which more is true than what is captured by the axioms.

For when faced with undecidability results, intuitive or heuristic thinking connected with an intended interpretation can also suggest additional axioms, or at least, that has been one guiding idea in foundations†. The heuristic notion of *set* is, of course, the notion of one thing made of many things by "collection into a whole." The heuristic picture of the *set-theoretic universe*, found in writings of Zermelo from the 1930s and Gödel 1947, is of what is obtained by starting from whatever non-sets or *Urelemente*, if any, that one wants to consider, and then successively forming sets of things one has so far. Even with *zero* Urelemente—the case of "pure" sets, to which set-theorists typically confine their attention nowadays—this first gives the empty set $\varnothing$, the set of all things one has when one does not yet have anything. It then gives $\{\varnothing\}$, and then gives the Zermelo set $\{\{\varnothing\}\}$ and the Von Neumann set $\{\varnothing, \{\varnothing\}\}$. Now having four items there are $2^4 = 16$ sets of them, of which $16 - 4 = 12$ are new, and then $2^{16} = 65{,}536$ items, of which $65{,}520 = 65{,}536 - 16$ are new, and so on. And after all finite levels there are sets like

$$\{\varnothing, \{\varnothing\}, \{\{\varnothing\}\}, \ldots\}$$

to give us a level ω, and on into the transfinite. This is the "iterative conception" or "cumulative hierarchy."

The basic picture is supplemented by thoughts of *maximality*: the hierarchy is to be as *wide* as possible, allowing at each new level absolutely *arbitrary* sets of items we have so far. (In particular, the thought is that there is absolutely no requirement whatsoever of *definability*, such as would have been imposed by opponents of AC.) There is also the thought that the hierarchy is to be as *high* as possible, allowing *arbitrary* levels. Large-cardinal axioms (which may be considered to begin with replacement, which is needed to get to level ω + ω) are motivated as partial expressions of this idea of *maximality*.

There seems to be a limit, however, to how far known large cardinal axioms explored by foundations* can be viewed as intrinsically necessary or conceptually implicit in the heuristic picture of the set-theoretic universe. Reflection principles (beginning, but only beginning with the theorems of ZFC going by that name) embody the idea that the absolutely infinite is indescribably large, so that any description we offer of it is an understatement, true already of the large enough transfinite. Such principles are arguably direct expressions of the maximal height idea, and hence conceptually implicit in the heuristic picture behind the existing formal axioms, and good candidates for addition as further axioms with a claim to intrinsic necessity. Peter Koellner (2009), however, has studied the matter closely and concluded that reflection principles so motivated won't take us very high up the scale (dropping us somewhere between weakly compact and measurable cardinals).

Maddy (1988) has studied the "rules of thumb" suggesting much larger large cardinals, but it is observable if one looks at all critically at them that many of these pretty clearly cannot be construed (nor does Maddy attempt to construe them) as simply making explicit what was already implicit in the heuristic maximal height idea. The larger of the large cardinals seem to have become objects of study more on account of the attractiveness of their consequences (such as the implications of Woodin cardinals for descriptive set theory) than from any sense of intrinsic necessity. But the known consequences, attractive as they are in their bearing on descriptive set theory, seem still to leave us far from any resolution of the continuum problem.

This is one ground for the pessimism of the defeatists, who are ready at this stage to abandon foundations†, and any hope of finding an answer to the continuum problem that could be considered the intrinsically necessary answer and not merely an aesthetically pleasing answer. It should be noted that however strong such grounds for pessimism may seem, *such* grounds do not rest on simple rejection of any notion of an "intended interpretation" having more in it that what is captured by the conventional axioms of ZFC. It can coherently be accepted that there is enough of an "intended interpretation" to motivate adding at least some large cardinal axioms going beyond ZFC, while denying that there is enough to motivate any further axiom that would settle CH one way or the other. Or at least, the incoherence of such a position has not been demonstrated.

Insofar as structuralist ideas, which *do* reject any notion of "intended interpretation," are relevant, they would at first glance appear to motivate a pessimism beginning at a much earlier stage, a pessimistic or rather a rejectionist attitude towards the whole project of foundations†, and not just a lack of optimism about how far such a project can be successfully carried out. The kind of pessimism that arises for some working within foundations† is a kind of pessimism that sees the road that has been followed beyond the formal axioms, appealing to a heuristic picture of an intended interpretation behind those axioms, as petering out at a certain point. By contrast, the structuralist kind of pessimism might be accused by practitioners of foundations† of "blocking the road of inquiry" from the beginning with its negativity about intended interpretations, forbidding us even to follow and *see* where the road gives out, if it eventually does.

It is characteristic, however, of philosophical theses at a high level of generality that while they may seem to be more congenial to one rather than another of various positions taken on various more specific issues, in the end they turn out to be susceptible to interpretations or implementations compatible with just about any position on those more specific issues. And so it may be with structuralism.

In the case of the natural numbers, Parsons 2008, even while holding to structuralist ideas, and denying that we have any "intuition" of the natural numbers as such in any philosophically interesting sense, nonetheless maintains that Hilbert was right to claim that we *do* have a kind of Kantian, spatiotemporal intuition of *one particular* progression, that given by the stroke numerals:

$$|, ||, |||, ||||, \ldots$$

Because all progressions are isomorphic, any structural property we can discover intuitively in the case of the Hilbert model of stroke numerals will apply to every progression, and to the natural numbers. Thus intuition and an "intended interpretation" *can* play a role even for an avowed structuralist, at least at the level of arithmetic.

In the case of the real numbers, similarly, even while holding to structuralist ideas, and denying that we have any "intuition" of the real numbers in any philosophically interesting sense, one could nonetheless maintain that we *do* have a kind of Kantian, spatiotemporal intuition of *one particular* complete ordered field, that formed by ratios of lengths in the Euclidean plane. *Because all complete ordered fields are isomorphic,* any structural property we can discover intuitively in the case of the Euclidean model of length-ratios will apply to every complete ordered field, and to the real numbers. Thus again intuition and an "intended interpretation" perhaps *can* play a role even for an avowed structuralist also at the level of analysis.

What about set theory? If we speak of "intuition" here it cannot be in a Kantian, spatiotemporal sense; but it is not clear that this should matter if we do have clear intuitions of some other kind of the Zermelo-Gödel model of a maximal cumulative hierarchy of "collections into a whole." The main problem seems rather to be that, while in arithmetic we have a rigorously defined notion of progression, for which it can be proved that all such are isomorphic, and while in analysis we have a rigorously defined notion of complete ordered field, for which it can be proved that all such are isomorphic, we do not have in set theory any rigorously defined notion of "maximal cumulative hierarchy," for which it can be proved that all such are isomorphic.

In technical terms used by logicians, second-order arithmetic and second-order analysis are by theorems of Dedekind "categorical," with all models isomorphic, while for second-order set theory (concerning only "pure sets" with zero *Urelemente*) we have only the partial result that any two models of the same "height" are by a theorem of Zermelo isomorphic and so have the same "width." This is enough to imply that there is an objectively "right" answer to the continuum question, though without any intuition of any specific model, comparable to Hilbert stroke-numeral or Euclidean length-ratio models in arithmetic or analysis, we are left without any obvious strategy for discovering what that "right" answer *is*.

In any case, it is not clear how many structuralists accept even this much categoricity, and are prepared to grant that, even without a rigorous definition, we have some kind of grasp on the notion of a "maximal cumulative hierarchy," unique up to isomorphism.[22] Most structuralists who consider the question, including notably Parsons, seem to hold, on the contrary, not only that there is no unique "intended interpretation" or "preferred model" of set theory, but also that we cannot speak even of an intended or preferred *isomorphism type* of interpretation or model in set theory, comparable to the progressions in arithmetic and the complete ordered fields in analysis. But while it may be that most structuralists in practice do take a line "blocking the road of inquiry" early on, I know of no fully worked-out argument why all structuralists in principle *must* take such a line. Insofar as this is so, there are aspects of the relationship between structuralism and foundations that have not yet been fully worked out. And thus my exploration of the notion of structure must, like my exploration of the notion of rigor, end by leaving certain questions for the future.

---

[22] If I understand the discussion of Gödel in Martin 2005 aright, the position that Martin attributes to Gödel would seem to be of this kind. But my attempt to keep the exposition at a semi-popular level had the consequence, as we have entered deeper and deeper into technicalities, that my account has become sketchier and sketchier. For the reader who would like a clearer, more detailed, and fuller account, there is no help for it but to turn to somewhat more technical treatments. I have already cited the most relevant works by structuralists. As for work by "optimists" about foundations†, the place to begin is with Gödel 1947, after which one can turn to the subsequent literature it has inspired, perhaps beginning with the essays in Kennedy 2014.

Bibliography

Appel, Kenneth and Wolfgang Haken (1989) *Every Planar Map is Four-Colorable*, Providence, RI: American Mathematical Society.

Arnold, Vladimir Igorevich, Michael Atiyah, Peter Lax, and Barry Mazur (eds) (2000) *Mathematics, Frontiers and Perspectives*, Providence, RI: American Mathematical Society.

Artin, Michael, Alexandre Grothendieck, and Jean-Louis Verdier (eds) (1972) *Séminaire de Géométrie Algebrique du Bois-Marie, 1963–1964: Théorie des topos et cohomologie étale des schémas* (SGA 4), 3 vols, Springer Lecture Notes in Mathematics, 269, 270, and 205, Berlin: Spinger-Verlag.

Asimov, Isaac (1971) *Isaac Asimov's Treasury of Humor: A Lifetime Collection of Favorite Jokes, Anecdotes, and Limericks with Copious Notes on How to Tell Them and Why*, Boston: Houghton Mifflin.

Atiyah, Michael (ed.) (1994) Responses to "Theoretical Mathematics: Towards a Cultural Synthesis of Mathematics and Theoretical Physics," by A. Jaffe and F. Quinn. *Bulletin of the American Mathematical Society*, 30: 178–207.

Awodey, Steve (2004) An Answer to Hellman's Question: "Does Category Theory Provide a Framework for Mathematical Structuralism?." *Philosophia Mathematica*, 4: 209–37.

Bell, Eric Temple (1956) The Prince of Mathematicians. In Newman 1956c, i. 295–339.

Benacerraf, Paul (1965) What Numbers Could Not Be. *Philosophical Review*, 74: 47–73 (reprinted in Benacerraf and Putnam 1983, 272–94).

Benacerraf, Paul (1973) Mathematical Truth. *Journal of Philosophy*, 70: 661–79 (reprinted in Benacerraf and Putnam 1983, 403–20).

Benacerraf, Paul and Hilary Putnam (1983) *Philosophy of Mathematics: Selected Readings*, 2nd edn, Englewood Cliffs, NJ: Prentice-Hall.

Berndt, Bruce Carl (1998) An Overview of Ramanujan's Notebooks. In Butzer 1998, 119–46.

Beth, Evert Willem and Jean Piaget (1974) *Mathematical Epistemology and Psychology*, Dordrecht: Reidel.

Beth, Evert Willem, Hendrik Josephus Pos, and J. H. A. Hollak (Jan) (eds) (1949) *Proceedings of the 10th International Congress of Philosophy, Amsterdam 1948*, Amsterdam: North Holland.

Birkhoff, Garrett and Saunders Mac Lane (1941) *A Survey of Modern Algebra*, New York: Macmillan.

Blanc, Georges and Anne Preller (1975) Lawvere's Basic Theory of the Category of Categories. *Journal of Symbolic Logic*, 40: 14–18.

Boolos, George Stephen (1971) The Iterative Conception of Set. *Journal of Philosophy*, 68: 215–31 (reprinted in Benacerraf and Putnam 1983, 486–502).

Bourbaki, Nicolas [collective pseud.] (1939) *Théorie d'ensembles: Fascicule de résultats*, Paris: Hermann.

Bourbaki, Nicolas [collective pseud.] (1949) Foundations of Mathematics for the Working Mathematician. *Journal of Symbolic Logic*, 14: 1–8.

Bourbaki, Nicolas [collective pseud.] (1950) The Architecture of Mathematics. *American Mathematical Monthly*, 57: 221–32.

Bourbaki, Nicolas [collective pseud.] (1957) *Théorie d'ensembles*, Paris: Hermann.

Brouwer, Luitzen Egbertus Jan (1949) Consciousness, Philosophy, and Mathematics. In Beth et al. 1949, iii. 1235–49 (reprinted in Benacerraf and Putnam 1983, 90–6).

Browder, Felix Earl (ed.) (1974) *Mathematical Developments Arising Out of Hilbert Problems, Proceedings of Symposia in Pure Mathematics, 28/2,* Providence, RI: American Mathematical Society.

Burgess, John Patton (1977) Forcing. In K. J. Barwise (ed.), *Handbook of Mathematical Logic*, Amsterdam: North Holland, 403–52.

Burgess, John Patton (2005) *Fixing Frege*, Princeton: Princeton University Press.

Burgess, John Patton and Gideon Rosen (1997) *A Subject with No Object: Strategies for Nominalistic Reinterpretation of Mathematics*, Oxford: Oxford University Press.

Butzer, Paul Leo, Hubertus Theodor Jongen, and Walter Oberschelp (eds) (1998) *Charlemagne and his Heritage: 1200 Years of Civilization and Science in Europe,* ii. *Mathematical Arts*, Turnhout: Brepols.

Butts, Robert and Jaakko Hintikka (eds) (1977) *Logic, Foundations of Mathematics and Computability Theory: Proceedings of the 5th International Congress of Logic, Methodology, and Philosophy of Science, London, Ontario 1975,* i, Reidel: Dordrecht.

Cantor, Georg (1915) *Contributions to the Founding of the Theory of Transfinite Numbers*, tr. from the German by Philip E. B. Jourdain, Chicago: Open Court (also available in 1952 Dover reprint).

Carlson, James, Arthur Jaffe, and Andrew Wiles (eds) (2006) *The Millennium Prize Problems*, Boston: Clay Mathematics Institute, in collaboration with the American Mathematical Society.

Carr, George Shoobridge (1886) *A Synopsis of Elementary Results in Pure Mathematics: Containing Propositions, Formulæ, and Methods of Analysis, with Abridged Demonstrations*, Cambridge: Macmillan & Bowes.

Carroll, Lewis [pseud. Charles Lutwidge Dodgson] (1897) *Symbolic Logic*, London: Macmillan (many reprints available).

Chihara, Charles S. (2003) *A Structural Account of Mathematics*, Oxford: Oxford University Press.

Chong, Chi-Tat, Qi Feng, Theodore Allen Slaman, and Hugh Woodin (eds) (2014) *Infinity and Truth*, National University of Singapore Institute for Mathematical Sciences Lecture Notes Series, 25, Hackensack: World Scientific.

Colyvan, Mark (2014) Indispensability Arguments in Philosophy of Mathematics. In Edward N. Zalta (ed.), *The Stanford Encyclopedia of Philosophy* (Spring 2014 edn), <http://plato.stanford.edu/archives/spr2014/entries/mathphil-indis>.

Conway, John Horton (1976) *On Numbers and Games*, New York: Academic Press.

Cooke, Harold Percy and Hugh Tredennick (tr. and ed.) (1938) *Aristotle, I: Categories, On Interpretation, Prior Analytics*, Greek text with facing English translation, Loeb Classical Library, 325, Cambridge, MA: Harvard University Press.

Dedekind, Richard (1901) *Essays on the Theory of Numbers, I. Continuity and Irrational Numbers, II. The Nature and Meaning of Numbers*, tr. from the German by Wooster Woodruff Beman, Chicago: Open Court.

Dieudonné, Jean (1982) *A Panorama of Pure Mathematics: As Seen by N. Bourbaki*, tr. from the French by I. G. Macdonald, New York: Academic Press.

Dummett, Michael Anthony Eardley (1975) The Philosophical Basis of Intuitionistic Logic. In Shepherdson and Rose 1975, 5–40 (reprinted in Benacerraf and Putnam 1983, 97–129).

Eilenberg, Samuel and Saunders MacLane [later Mac Lane] (1945) General Theory of Natural Equivalences. *Transactions of the American Mathematical Society*, 58: 231–94.

Eilenberg, Samuel, David Kent Harrison, Saunders MacLane [later Mac Lane], and Helmut Röhrl (eds) (1966) *Proceedings of the Conference on Categorical Algebra, La Jolla 1965*, Berlin: Springer-Verlag.

Einstein, Albert (1949) Remarks Concerning the Essays Brought Together in this Co-operative Volume. In Schilpp 1949, 665–88.

Feferman, Solomon (1977) Categorical Foundations and Foundations of Category Theory. In Butts and Hintikka 1977, 149–69.

Feferman, Solomon (2013) Foundations of Unlimited Category Theory: What Remains to be Done. *Review of Symbolic Logic*, 6: 6–15.

Field, Hartry H. (1980) *Science without Numbers*, Princeton: Princeton University Press.

Frege, Gottlob (1983) The Concept of Number. Excerpted from *The Foundations of Arithmetic*, tr. from the German by Michael S. Mahoney (reprinted in Benacerraf and Putnam 1983, 130–59).

Gabbay, Dov M., Akihiro Kanamori, and John Woods (eds) (2012) *Handbook of the History of Logic*, vi. *Sets and Extensions in the Twentieth Century*, Amsterdam: North-Holland (Elsevier).

Gerhardt, Carl [Karl] Immanuel (ed.) (1859) *Leibnizens mathematische Schriften*, Halle: Verlag von H. W. Schmidt.

Gödel, Kurt (1947) What is Cantor's Continuum Problem? *American Mathematical Monthly*, 9: 515–25 (reprinted with additions in Benacerraf and Putnam 1983, 470–85).

Gonthier, Georges (2008) Formal Proof: The Four-Color Theorem. *Notices of the American Mathematical Society*, 55: 1382–93.

Gowers, William Timothy (2000) The Two Cultures of Mathematics. In Arnold et al. 2000, 65–78.

Gray, Jeremy (1991) Did Poincaré Say "Set Theory is a Disease"? *Mathematical Intelligencer*, 13: 19–22.

Haaparanta, Leila (2008) *The Development of Modern Logic*, Oxford: Oxford University Press.

Hadamard, Jacques (1945) *An Essay on the Psychology of Invention in the Mathematical Field*, Princeton: Princeton University Press.

Hahn, Hans (1956a) The Crisis in Intuition. Anonymously tr. from the German, in Newman 1956c, iii. 956–77.

Hahn, Hans (1956b) Infinity. Anonymously tr. from the German, in Newman 1956c, iii. 1593–1611.

Hallett, Michael (1982) *Cantorian Set Theory and Limitation of Size*, Oxford Logic Guides, Oxford: Oxford University Press.

Hamilton, William Rowan (1837) Theory of Conjugate Functions, or Algebraic Couples; with a Preliminary and Elementary Essay on Algebra as the Science of Pure Time. *Transactions of the Royal Irish Academy*, 17: 293–422.

Hardy, Godfrey Harold (1908) *A Course of Pure Mathematics*, Cambridge: Cambridge University Press (2nd edn, Cambridge: Cambridge University Press, 1914).

Heath, Thomas Little (tr. and ed.) (1912) *The Method of Archimedes: Recently Discovered by Heiberg: A Supplement to* The Works of Archimedes *1897*, Cambridge: Cambridge University Press.

Heath, Thomas Little (tr. and ed.) (1926) *The Thirteen Books of Euclid's Elements: Translated from the Text of Heiberg, with Introduction and Commentary*, 2nd edn, 3 vols, New York: Macmillan (available also in Dover reprint).

Hellman, Geoffrey (1993) *Mathematics without Numbers: Towards a Modal-Structural Interpretation*, Oxford: Oxford University Press.

Hellman, Geoffrey (2001) Three Varieties of Mathematical Structuralism. *Philosophia Mathematica*, 9: 184–211.

Hellman, Geoffrey (2003) Does Category Theory Provide a Framework for Mathematical Structuralism? *Philosophia Mathematica*, 11: 129–57.

Hilbert, David (1902) *The Foundations of Geometry*, tr. from the German by E. J. Townsend, LaSalle, IL: Open Court.

Hilbert, David (1983) On the Infinite. Tr. from the German by Erna Putnam and Gerald J. Massey (reprinted in Benacerraf and Putnam 1983, 183–201).

Hilbert, David and Stephan Cohn-Vossen (1952) *Geometry and the Imagination*, tr. from the German by P. Nemenyi, New York: Chelsea.

Hodges, Wilfrid (2008) Set Theory, Model Theory, and Computability Theory. In Haaparanta 2008, 471–98.

Horgan, John (1993) The Death of Proof. *Scientific American* (Oct.): 92–103.

Huffman, Carl (2014) Pythagoras. In E. Zalta (ed.), *Stanford Encyclopedia of Philosophy*, Summer 2014 edn, <http://plato.stanford.edu/archives/sum2014/entries/pythagoras>.

Hume, David (1739) *A Treatise of Human Nature: Being an Attempt to Introduce the Experimental Method of Reasoning into Moral Subjects*, London: John Noon (also available in many subsequent edns).

Isbell, John Rolfe (1967) Review of Lawvere 1966. *Mathematical Reviews*, 34: 7332.

Jaffe, Arthur and Frank Quinn (1993) "Theoretical Mathematics": Towards a Cultural Synthesis of Mathematics and Theoretical Physics. *Bulletin of the American Mathematical Society*, 29: 1–13.

Jesseph, Douglas M. (1993) *Berkeley's Philosophy of Mathematics*, Chicago: University of Chicago Press.

Kalai, Gil (2008) Can Category Theory Serve as the Foundation of Mathematics? Entry in blog *Combinatorics and More*, 4 Dec. <http://gilkalai. wordpress. com/2008/12/04/can-category-theory-serve-as-the-foundation-of-mathematics>.

Kanamori, Akihiro (2012) In Praise of Replacement. *Bulletin of Symbolic Logic*, 18: 45–90.

Keisler, Howard Jerome (2000) *Elementary Calculus: An Infinitesimal Approach*, on-line edn, <http://www.math.wisc.edu/~keisler/calc.html>.

Kennedy, Juliette (ed.) (2014) *Interpreting Gödel: Critical Essays*, Cambridge: Cambridge University Press.

Kennedy, Juliette and Roman Kossak (2011) *Set Theory, Arithmetic, and Foundations of Mathematics: Theories, Philosophies*, Lecture Notes in Logic, 36, Cambridge: Cambridge University Press.

Kleiman, Steven Lawrence (1974) Problem 15. Rigorous Foundation of Schubert's Enumerative Calculus. In Browder 1974, 445–82.

Kline, Morris (1972) *Mathematical Thought from Ancient to Modern Times*, Oxford: Oxford University Press.

Koellner, Peter (2009) On Reflection Principles. *Annals of Pure and Applied Logic*, 157: 206–19.

Krantz, Steven George (1994) The Immortality of Proof. *Notices of the American Mathematical Society*, 41: 10–13.

Krantz, Steven George (2011) *The Proof is in the Pudding: The Changing Nature of Mathematical Proof*, New York: Springer-Verlag.

Lakatos, Imre (ed.) (1967) *Problems in the Philosophy of Mathematics*, Amsterdam: North Holland.

Lakatos, Imre (1976) *Proofs and Refutations: The Logic of Mathematical Discovery*, Cambridge: Cambridge University Press.

Lawvere, William (1966) The Category of Categories as a Foundation for Mathematics. In Eilenberg et al. 1966, 1–21.

Lawvere, William (2005) An Elementary Theory of the Category of Sets (Long Version) with Commentary (by Colin McLarty and Author). *Reprints in Theory and Application of Categories*, 12: 1–35.

Lebesgue, Henri (1922) *Notice sur les travaux scientifiques de M. Henri Lebesgue*, Toulouse: Édouard Privat.

Leibniz, Gottfried Wilhelm (1701) Mémoire de Mr. G. G. [Monsieur Gottfried Guillaume] Leibniz touchant son sentiment sur le calculus différentiel (reprinted in Gerhardt 1859, v. 350).

Mac Lane, Saunders (1971) *Categories for the Working Mathematician*, Graduate Texts in Mathematics, Berlin: Springer-Verlag.

Mac Lane, Saunders and Garrett Birkhoff (1986) *Mathematics: Form and Function*, Berlin: Springer-Verlag.

Mac Lane, Saunders and Garrett Birkhoff (1988) *Algebra*, 3rd edn, New York: Chelsea.

McLarty, Colin (1991) Axiomatizing a Category of Categories. *Journal of Symbolic Logic*, 56: 1243–60.

McLarty, Colin (2004) Exploring Categorical Structuralism. *Philosophia Mathematica*, 12: 37–53.

McLarty, Colin (2005) Learning from Questions on Categorical Foundations. *Philosophia Mathematica*, 13: 44–60.

McLarty, Colin (2007) The Last Mathematician from Hilbert's Göttingen: Saunders Mac Lane as a Philosopher of Mathematics. *British Journal for the Philosophy of Science*, 58: 77–112.

McLarty, Colin (2010) What Does it Take to Prove Fermat's Last Theorem? Grothendieck and the Logic of Number Theory. *Bulletin of Symbolic Logic*, 16: 359–77.

Maddy, Penelope (1988) Believing the Axioms. *Journal of Symbolic Logic*, 53: 481–511 and 736–64.

Maddy, Penelope (1993) Wittgenstein's Anti-Philosophy of Mathematics. In Puhl 1993, 52–72.

Maddy, Penelope (1997) *Naturalism in Mathematics*, Oxford: Oxford University Press.

Mandelbrot, Benoît (1982) A Crisis in Intuition as Viewed by Felix Klein and Hans Hahn and its Resolution by Fractal Geometry. Unpublished manuscript, intermittently available online, excerpts appearing in Mandelbrot 1983.

Mandelbrot, Benoît (1983) *The Fractal Geometry of Nature: Updated and Augmented*, New York: W. H. Freeman.

Marquis, Jean-Pierre and Gonzalo Edmundo Reyes (2012) History of Categorical Logic: 1963–1977. In Gabbay et al. 2012, 689–800.

Martin, Donald Anthony (2005) Gödel's Conceptual Realism. *Bulletin of Symbolic Logic*, 11: 207–24.

Mathias, Adrian Richard David (1992) The Ignorance of Bourbaki. *Mathematical Intelligencer*, 3: 13–14.

Mathias, Adrian Richard David (2001) The Strength of Mac Lane Set Theory. *Annals of Pure and Applied Logic*, 101: 107–234.

Mathias, Adrian Richard David (2014) Hilbert, Bourbaki, and the Scorning of Logic. In Chong et al. 2014, 47–56.

Moore, Gregory H. (1982) *Zermelo's Axiom of Choice: Its Origins, Development, and Influence*, New York: Springer-Verlag.

Newman, James Roy (1956a) The Rhind Papyrus. In Newman 1956c, i. 170–9 (reprinted from *Scientific American* 1952).

Newman, James Roy (1956b) Srinivasa Ramanujan. In Newman 1956c, i. 368–80 (reprinted from *Scientific American* 1948).

Newman, James Roy (1956c) *The World of Mathematics: A Small Library of the Literature of Mathematics from A'h-mosé the Scribe to Albert Einstein, Presented with Commentaries and Notes*, 4 vols, New York: Simon & Schuster.

Oxtoby, John C. (1980) *Measure and Category*, 2nd edn, New York: Springer-Verlag.

Pais, Abraham (1982) *"Subtle is the Lord . . . ": The Life and Science of Albert Einstein*, Oxford: Oxford University Press.

Parsons, Charles (1965) Frege's Theory of Numbers. In Max Black (ed.), *Philosophy in America*, Ithaca, NY: Cornell University Press, 180–203.

Parsons, Charles (1990) The Structuralist View of Mathematical Objects. *Synthese*, 84: 303–46.

Parsons, Charles (2008) *Mathematical Thought and Its Objects*, Cambridge: Cambridge University Press.

Peano, Giuseppe (1901) *Formulaire de mathématiques*, Paris: Gauthier-Villars.

Poincaré, Henri (1920) *Science et méthode*, Paris: Flammarion.

Polya, George (1945) *How to Solve it: A New Aspect of Mathematical Method*, Princeton: Princeton University Press (partially reprinted in Newman 1956c, iii. 1980–93).

Polya, George (1954) *Mathematics and Plausible Reasoning*, i. *Induction and Analogy in Mathematics*, ii. *Patterns of Plausible Inference*, Princeton: Princeton University Press.

Puhl, Klaus (ed.) (1993) *Wittgenstein's Philosophy of Mathematics*, Vienna: Hölder-Pichler-Tempsky.

Quine, Willard Van Orman (1937) New Foundations for Mathematical Logic. *American Mathematical Monthly*, 44: 70–80.

Quine, Willard Van Orman (1951) *Mathematical Logic*, revised edn, Cambridge, MA: Harvard University Press.

Quinn, Frank (2012) A Revolution in Mathematics? What Really Happened a Century Ago, and Why It Matters Today. *Notices of the American Mathematical Society*, 59: 31–7.

Rademacher, Hans and Otto Toeplitz (1966) *The Enjoyment of Mathematics: Selections from Mathematics for the Amateur*, tr. from the German by Herbert Zuckerman, Princeton: Princeton University Press; reprint New York: Dover, 1990.

Reid, Constance (1970) *Hilbert*, New York: Springer-Verlag.

Reimer, David (2014) *Count like an Egyptian: A Hands-On Introduction to Ancient Mathematics*, Princeton: Princeton University Press.

Resnik, Michael (1997) *Mathematics as a Science of Patterns*, Oxford: Oxford University Press.

Robinson, Abraham (1966) *Non-Standard Analysis*, Amsterdam: North Holland.

Robinson, Abraham (1967) The Metaphysics of the Calculus. In Lakatos 1967, 28–46.

Rosen, Gideon and John Patton Burgess (2005) Nominalism Reconsidered. In S. Shapiro (ed.), *Handbook of Philosophy of Mathematics and Logic*, 460–82. Oxford: Oxford University Press.

Russell, Bertrand Arthur William (1919) *Introduction to Mathematical Philosophy*, London: George Allen & Unwin (selections reprinted in Benacerraf and Putnam 1983, 160–83).

Russell, Bertrand Arthur William (1946) *History of Western Philosophy: And Its Connection with Political and Social Circumstances from the Earliest Times to the Present Day*, London: George Allen & Unwin.

Russell, Bertrand Arthur William (1956) Mathematics and the Metaphysicians. In Newman 1956c, iii. 1576–90 (reprinted from *Mysticism and Logic*, 1929).

Schilpp, Paul Arthur (ed.) (1949) *Albert Einstein: Philosopher-Scientist*, Library of Living Philosophers, 7, Evanston, IL: Library of Living Philosophers.

Shapiro, Stewart (1997) *Philosophy of Mathematics: Structure and Ontology*, Oxford: Oxford University Press.

Shepherdson, John C. and Harvey E. Rose (eds) (1975) *Logic Colloquium '73*, Amsterdam: North Holland.

Smith, David Eugene (ed.) (1929) *Source Book in Mathematics*, New York: McGraw-Hill.

Steiner, Mark (1975) *Mathematical Knowledge*, Ithaca, NY: Cornell University Press.

Tao, Terence (2013) *Compactness and Contradiction*, Providence, RI: American Mathematical Society.

Thomas, Ivor (tr. and ed.) (1939) *Greek Mathematical Works*, i. *Thales to Euclid*, Greek text with facing English translation, Loeb Classical Library, 335, Cambridge, MA: Harvard University Press.

Thurston, William Paul (1994) On Proof and Progress in Mathematics. *Bulletin of the American Mathematical Society*, 30: 161–77.

Tredennick, Hugh, and Edward S. Forster (tr. and ed.) (1960) *Aristotle II: Posterior Analytics, Topica*, Greek text with facing English translation, Loeb Classical Library, 391, Cambridge, MA: Harvard University Press.

Troelstra, Anne Sjerp (2011) History of Constructivism in the 20th Century. In Kennedy and Kossak 2011, 150–79.

Weber, Heinrich Martin (1893) Leopold Kronecker. *Mathematische Annalen*, 43: 1–25.

Weyl, Hermann (1949) *Philosophy of Mathematics and Natural Science*, Princeton: Princeton University Press.

Whitehead, Alfred North (1925) *Science and the Modern World: The Lowell Lectures, 1925*, New York: Macmillan.

Wittgenstein, Ludwig (1978) *Remarks on the Foundations of Mathematics*, revised edn, tr. from the German by G. E. M. Anscombe, Oxford: Basil Blackwell.

Zeilberger, Doron (1994) Theorems for a Price: Tomorrow's Semi-Rigorous Mathematical Culture. *Notices of the American Mathematical Society*, 40: 978–81.

Zeilberger, Doron (2014) Dr. Z's Opinions. <http://www.math.rutgers.edu/~zeilberg/OPINIONS.html>.

Index

Printed and bound by CPI Group (UK) Ltd, Croydon, CR0 4YY